PRAISE FOR *THE FARM IS HERE*

Agriculture is a key piece to solving the climate and ecological crisis. *The Farm is Here* is an honest look into how the humble act of regenerative organic farming may be our best bet to heal the planet—and ourselves.
— **Yvon Chouinard, Founder, *Patagonia***

The Farm is Here is a powerful testament to how deeply our lives are interwoven with the soil beneath our feet. With clarity and conviction, Jeff Tkach shows us that regenerative organic farming is not only about healing the land, it's about healing ourselves, our communities, and our planet. This book is both an invitation and a roadmap for the regenerative future we know is possible, grounded in ancient wisdom and urgent truth.
— **David Bronner, Cosmic Engagement Officer, *Dr. Bronner's***

The Farm is Here is a call to consciousness. Jeff Tkach reminds us that soil is not simply the ground beneath our feet, but the living foundation of our health, our societies, and our future.
— **Davide Bollati, Chairman, Davines Group**

An important book rooted in hope for our planet.
— **Rose Marcario, former CEO of *Patagonia* and founding board member of the Regenerative Organic Alliance**

Jeff Tkach reveals a way of seeing our relationship with the land as an opportunity for hope and healing. It reads like a soulful guide with practical steps that connect the human spirit to the resilience of nature. Through his extraordinary humility, we find a glimmer of our own, and an offer to join the journey. This book is a love letter to life — to the soil, to the seasons, and to the possibility that we can grow something better together.
— **John Chester, Founder of *Apricot Lane Farms* and director of *The Biggest Little Farm***

Let me just say - Jeff nailed it. I heartily recommend this book.
— **Will Harris, *White Oak Pastures***

THE FARM IS HERE

JEFF TKACH

2026

Copyright © 2026 New Farm

All rights reserved, including the right to reproduce this book or portions thereof in any form whatsoever. Permission inquiries may be directed to thefarmishere@rodaleinstitute.org

Cover photo: Ramon Madrid
Cover design: Brent French

Published in North America.

Paperback ISBN: 979-8-9940550-3-8

First print edition: 2026

For everyone that has ever worked on the mission of Rodale Institute: past, present, and future.

CONTENTS

Foreword	iii
1. The Farm Speaks to Me	1
2. Farm Yourself	13
3. The Soil is Alive	27
4. Nature's Rhythm	39
5. Steaming Piles of Life	49
6. The Guts of It	57
7. A Box of Vegetables?	69
8. The Truth About Regenerative	81
9. The Long Game	107
10. So Many Dave's	123
Epilogue	141
Join the Movement	147
Notes	149
Acknowledgements	161
About the Author	163

FOREWORD

They call it the Rodale Institute.
And that's fine. And accurate.
It is, after all, an institute.
But in my experience it's more like a *vortex*.
The Rodale Vortex. I like the sound of that.

Here's how it works:
First, you hear about the Rodale Institute and how
Aren't they the ones who coined the term ORGANIC farming?
or something like that, and you think that's cool because, well, who wants MORE chemicals in their food?

But then you meet someone from Rodale - in my case Jeff - and you're struck with their passion and intelligence and resilience and you quickly realize that these people are way out ahead and have been for a long, long time. You learn that what they're doing is way, way bigger than just

NOT SPRAYING TOXIC THINGS THAT KILL PEOPLE ON PLANTS

It's about an entirely different way of farming altogether that's better for our food and our bodies, better for the earth, and of course, better for the farmers themselves.
Well, you think, this sounds quite revolutionary.

Which it is, but you're just getting started.

Because once you're in the RODALE VORTEX you start hearing more and more stories and seeing more and more data while you're discovering that this isn't just about soil. It's about ecosystems and weather patterns and economics and what we mean by the word COST and farmers and farming and of course politics and sustainability and what we wear and the health of the entire earth and everybody on it.

Why isn't everybody talking about this?
You think to yourself.

These people have vital answers to so many of the questions so many people are asking about our future here together on earth.

That's the truth that keeps coming back to you. You're struck with how obvious it all is. How it's a massive undertaking with impossible odds and the clock is ticking and future generations are depending on it and yet...

It's doable.
Entirely doable.
Now.
That's what you keep experiencing with Rodale.
THIS CAN BE DONE.
And now you're in.
Way in.
So yes, it's about the soil and the science and the food.

And it's also about your heart.

How can I help?
What can we do?
How do we get the word out?
You're so thrilled to take the next step...
How could you not?
Welcome to the Rodale vortex.

—Rob Bell

THE FARM IS HERE

Rodale Institute corporate office

Chapter 1

THE FARM SPEAKS TO ME

JEFF TKACH

THE FARM IS HERE

The Farm speaks to me.

Sometimes it's loud. Like really loud. Especially in July.

My office is in a 250-year-old farmhouse.

Like other CEOs, I have a desk, a conference table, a laptop, and a phone. But when I take one step outside of my office door, I enter a 400-acre farm.

This farm is teeming with life.

Several times per day, I go for walks on the farm. The farm calls me to walk, and at some point in my day, I always answer. It is the moment in my day that I most cherish.

In the heat of early July, there's a *hummmmmm* coming from the ground. A pulsation. Proclaiming *I'm alive, I'm alive, I'm alive.*

Everywhere I look, nature is thriving. Apple orchards pumping fruit, amber waves of grain swaying in the breeze, vegetables pushing up from the earth, wildflowers splashing color onto the landscape. All growing in harmony, like a natural symphony.

I realize that this is a rare gift that very few people get to experience these days. Life on a farm.

We grow many different kinds of crops on our farm, while most farms focus on just a few crops like corn and soybeans.

Here on our farm, the land is alive with diversity, just as nature intended.

The beauty here isn't only in the plants; the animals are essential too.

The air is filled with birds singing. Some rare species have recently returned to our farm as if they know something significant is happening here.

We have pigs roaming freely in the pasture. We also have cows. Pigs and cows are not meant to live solely in buildings, which is how most livestock are raised these days.

I am not interested in a debate as to whether or not we should be consuming them. Animals simply make the land healthier. Research shows that they need the land to thrive, and the land needs them too.[1]

In one field, we let the meadow grow without mowing it in order to create a habitat for pollinators.

On some days, my job requires me to be on the farm late into the evenings. Around 9 p.m., millions of fireflies light up the night sky. It always stops me in my tracks. Like nothing I've seen before. Fireflies used to be prevalent. Until humans started spraying chemicals. But here, they're flourishing, thanks to a healthier ecosystem that we have created on our farm. Another sign that when we give the land what it needs, it gives back in ways we never could have imagined.

On a summer day, when I walk the farm and listen, I deepen my relationship with the life beneath my feet and all around me. But I also cherish connecting with the part of nature we often overlook: humanity.

As I walk down one of the pea-gravel paths on the farm, I see Dr. Arash Ghale, our director of research, checking in on one of our many research projects in progress. Our farm is a living laboratory for cutting-edge science.

Arash is from Iran. He was drawn to the organic farming movement as a young boy when his parents would take him on drives into the Iranian countryside. After attending college, he eventually got a job working for a German-based company that certified farms to the organic standards in Iran and throughout the Middle East.

His job was to go out and inspect farms to ensure they were compliant with international organic standards. He once inspected a farm near the volatile borders of Iran, Afghanistan, and Pakistan—an area so dangerous he needed an armed escort just to reach the farm fields.

For Arash, working at Rodale Institute is more than a job; it's the fulfillment of a long-held dream to be a part of a growing movement.

Everyone working here on this farm listens closely to the land, to the farmers, to the rhythm of the work — because that's where real intelligence lives. In an era of the acceleration of artificial intelligence, we turn to nature for the most intelligent answers.

At this moment, as I walk the farm, I feel connected to all the people on all the farms all over the planet. Farmers are some of the most innovative people in the world.

And then, there's this moment in July when I taste my first cherry tomato of the season. That first bite takes me back to being four years old in my great-grandmother's garden in Norristown, Pennsylvania. Her garden holds memories, too.

If I keep walking, I'll pass more of our fields growing a wide range of crops. Blackberries, kale, peas, purple potatoes, scallions, collard greens, green beans, lettuce, and the list goes on. Despite their differences, every fruit and vegetable is rooted in the same soil.

Just like us.

Our soil is more than dirt. It's expansive with life. If you took one teaspoon of soil from any of our fields and put it under a microscope, you'd find more than 10 billion microorganisms working together.[2] Think about it: That's more than the entire human population.

With that much life in a single teaspoon, imagine all that radiant energy circulating across 400 acres. That's a lot of life in one place.

The *hummmmmm* is the lifeforce that is pulsating across every square inch of this farm.

By the way, the average soil on an industrial farm is rapidly degrading.[3] Most farmland is turning into dirt, not living soil.

If we keep farming the way most farms do using industrial agricultural methods, we will face a soil health crisis in our lifetime. For real.

The good news?

It does not have to be this way. We can turn things around – quickly.

For as long as humans have been farming, natural, organic methods were the norm.

Indigenous people across the globe stewarded healthy ecosystems for thousands and thousands of years.[4] It's what kept the human species alive and evolving.

We need to recover the sacredness of the land that humans carefully cultivated for millennia. And we need modern science to understand how we can work in harmony with nature, using biological methods, to improve the health of the soil and to accelerate the healing of our ecosystems.

That's what we're doing here at Rodale Institute.

As J.I. Rodale, our founder, said in 1942,

"Healthy Soil = Healthy Food = Healthy People."®

The farm reminds me of that every day.

Soil health and human health are inextricably linked.

Sometimes, the farm isn't loud. It's quiet. Like really quiet.

January is the quietest month, if you ask me.

During the winter, the farm hands me a silent invitation to slow down.

An addiction to work can be cured by what the farm says in January. To rest the soil, to rest my body, and to rest *the* body—the Earth.

And yet, winter is one of the most mysteriously powerful times on the farm.

The land appears to be doing nothing, but in that stillness, *something* is happening.

Like a quiet that's actually quite purposeful.

A silence that's hidden from our eyes, taking place deep beneath the soil.

The snow plays a part in that. It offers the farm a blanket for protection and cover.

Some farmers even plant their seeds right before the first frost. The freezing pulls their seeds deeper into the ground, allowing them to germinate come springtime.

Sometimes growth comes from getting still and allowing things to happen beneath the surface.

The quiet of winter is a "nothing" that's actually a something.

Without this cold stillness, there would be no loud, rhythmic vibrancy in the warmth of summer.

You can't have July without January.

No matter the season, the farm is constantly teaching us.

Rodale Institute is part of this growing movement. It's happening here on our farm in Pennsylvania, on farms all around the world, in local communities, and in backyard gardens.

We're constantly learning more about how monumental this global shift back to the farm is and its implications for society. What it means for our future. And the expansive, abundant implications.

Take politics, for example. This movement is a wrecking ball for polarization.

If you work in regenerative organic agriculture, any division of a right and a left is demolished. We're all humans who need to eat healthy food. And the urban and rural divide gets smashed too. Most of us rely on rural communities to feed us.

Obviously, this extends to healthcare too, which often operates more like "sickcare."

In 1960, we spent twice as much on food as we did on healthcare. Sixty years later, we now spend three times more on healthcare than on food.[5]

And yet, we're collectively getting sicker. So, when we invest in a better food system, we feed people healthier food which makes them healthier and end up spending less on the outrageous costs of a broken healthcare system.

And of course, there's the chemicals.

For the first time in human history, people are eating food sprayed with significant amounts of chemicals. Many of which we are now discovering are linked to an increased risk of certain forms of cancer.[6] The chemical playbook of commercial agriculture is eradicating life in the soil in the name of extraction, efficiency, yields, and profit. But again, we know how to turn this around.

At the deepest level, our society is beginning to reconnect with where our food comes from, how what we're eating was produced, and ensuring everything we grow is nutrient rich.

Everything we do in regenerative organic agriculture is rooted in endless innovation combined with ancient wisdom. In our movement, there is no playbook, no chemicals, and no extraction. Simply put: Extraction is no longer sustainable.

Humans long for what is real and enduring.

In a world that is presumably more connected than ever through social media, we are lonelier than ever as a society. The farm has a lot to teach us about authentic community.

Annie Brown in Boulder, Colorado started a community potluck in partnership with her local farmers market called the Boulder County Seasonal Supper Club. Four times a year, they choose a vegetable that's in season and invite the community to make a dish that showcases that vegetable. What started as a few people making vegetable dishes grew into a sought-after event, gathering upwards of a hundred people together in the name of local, organic food.

Here at Rodale Institute, I see a similar phenomenon with our employee CSA. Yes, we give our employees a free share of vegetables each week to take home to their families. This is a component of our healthcare benefits package. We are investing in the health of our employees and their families, and in turn, we create a stronger and more connected workplace culture.

Further along in my walk, I see our students who have come from all over the world to spend nine months with us, learning how to become regenerative organic farmers.

Every Thursday at noon, the entire Rodale Institute staff gathers to collect produce from the farm grown by our students, which has now become a weekly community celebration and ritual.

We've built a workplace culture that literally breaks bread together.

Turns out, food and regenerative organic farming are the antidotes to loneliness. Creating a shared space for connection, for meeting new people, and for delivering nourishment that goes beyond the body and feeds the ache of the soul.

What we keep discovering is that the farm can transform nearly every area of our lives. The farm contains answers to our world's greatest problems.

What we're learning in this movement is the things we were told had nothing to do with each other are actually interconnected.

Everything from eating a proper tomato to getting good sleep to who you vote for to feeling like you belong, it's all the same thing, endlessly connected, like a giant infinity loop.

Labels, categories, and "us vs. them" become absurd when sharing a meal together from the farm. When we begin to reconnect with our food, we reconnect with each other.

This is the miracle of life on Earth.

What we put on our plate can change the future of humanity.

Really, the farm is a living, breathing organism that needs to be fed and nourished, just like you and me. When we feed the land, it feeds us.

And so, I walk and walk and walk the farm. And I listen for what it has to teach me today.

The farm tells me reciprocity, nourishment, and connection are the way forward. That we are all called to be farmers—stewards, caring for the land, ourselves, and each other.

So, this book is for everyone.

And for me, it's really, really personal.

Jeff Tkach, CEO, Rodale Institute

Chapter 2

FARM YOURSELF

THE FARM IS HERE

It started with a fever.

On a gorgeous Tuesday in October, the sky was bright blue, the temperature was exquisite, the trees were bursting with autumn colors, and I went home from work feeling very, very sick.

Earlier that year I had taken a new job. My wife Jackie and I had relocated from our life in Colorado to a rental home in Pennsylvania surrounded by a forest.

From afternoon to sunset, I sweated through two sets of sheets. I had no idea what was happening to me. Was this the flu?

Things kept getting worse and worse. I couldn't sleep. When I did, I had the weirdest dreams. Or more like nightmares. I felt sicker than I'd ever felt in my life.

Being sick wasn't new to me. As a young boy, I felt ill all the time. At four years old I was diagnosed with asthma and severe allergies. I'd go to the doctor, get medicine, and repeat this cycle over and over - for most of my childhood. Many of the medicines that my doctor prescribed made me feel worse.

By the time I turned 13, I had fallen into a state of depression and remember saying to myself, *I don't want to live this way anymore. I want to be healthy.*

A commercial came across the television for a magazine called *Men's Health*. The advertisement featured a guy with six-pack abs, a big smile, and a ton of confidence. I remember thinking how I wanted to be like that guy.

So I asked my mother to buy me a copy of Men's Health and I devoured it from cover to cover about nine times. Over and over again, I read that one copy of Men's Health magazine.

I thought to myself, Maybe if I do everything that this magazine tells me to do, including eating healthier foods and exercising like the guy in the magazine, maybe I will get healthy too.

I then gave my mom a grocery list of the foods that the magazine told me that I should eat if I wanted to be healthy.

I started to replace all of the processed foods that I was consuming (it was the late 1980s and ultra-processed and packaged food was booming) with foods that the magazine recommended for optimal health. And, little by little, I started to feel better. Over the next six months, I lost a lot of weight, got off all the medicines, and started playing sports.

From my earliest memories until I was 13, I struggled with my health. But then I felt empowered to be healthy. That experience as a 13-year-old boy shaped the trajectory of the rest of my life.

Fast-forward to that gorgeous day in October and the feeling of sickness washing over me.

I didn't have an established doctor yet in Pennsylvania. So I called the number on the back of my health insurance card.

I was sent to a group of Internal Medicine doctors working at the local hospital.

When I finally got in with one of the doctors, I told him everything that I was feeling.

He told me, "Stay home, get some rest, and drink lots of liquids. You will be fine."

So I followed the doctor's orders. Within a week, I tried to go back to work. For half a day.

I was going over numbers on spreadsheets in my office with one of my colleagues and I had to get up and leave the meeting because I could not focus.

For the next three months, I was trapped in an endless cycle of forcing myself to go to work, then crashing again. I was not getting better. And I was scared. Something was seriously wrong with me. I had never felt this sick in my entire life.

I should mention that the job that I had relocated from Colorado to Pennsylvania for was to be the managing director of the nation's leading health magazine, *Prevention,* with over three million readers and owned by Rodale Inc., the same company that also published *Men's Health*. Yes, I became the leader of a national health magazine, a career path that was likely motivated by my experience as a 13-year-old boy.

I was now 37 years old and was as physically fit as I'd ever been. I'd go on 100-mile bike rides. I had a sixpack of abs like the guy on TV.

And now I was sick. Really sick. And I wasn't getting better.

When I went back to see the doctor, he ran a bunch of tests. No answers or conclusive evidence for the cause of my sudden health collapse. Based on symptoms that I was exhibiting, my doctor prescribed a strong antibiotic in the hopes that it would target the vector or the root cause, but to no avail. If anything, the antibiotic made me feel worse.

That Christmas was a living hell. I spent most of the day in bed.

By now I'd seen six different medical specialists. Each of them were trying to help me to find answers to this mysterious illness. Not one of them was able to reach any conclusions and I didn't know what to do next.

So I went back to the first doctor.

This time, he convinced me I was having a mental breakdown. He wrote me a prescription for an antidepressant along with a letter recommending medical leave to my employer.

Everything was falling apart.

My body, my job, my life.

It was a deeply humbling, even humiliating, time of my life. I called my friend Paul who was the head of human resources for the company I worked for and I explained everything.

Paul was so wonderful at that moment. He told me to take as much time as I needed. He also made it clear that if I took medical leave, I wouldn't be allowed to work at all—not even to explain my health situation to my team. This is what the insurance company would mandate.

THE FARM IS HERE

I didn't go back to work in January. And I never found out what my team was told about me and my sudden leave of absence.

Jackie was struggling too, watching her partner fall apart. The stress of seeing me deteriorate put a strain on us.

We moved to a rented carriage house on a horse farm during that season. Most days, I laid on the couch and stared at the ceiling.

Anxiety and depression became my reality.

There were a few beautiful days that January. I'd sit in the sun, feeling nature wash over me. I made a new friend. A horse named Chance. I fed him carrots. He and I had conversations.

One day my friend Jeff, who was the head of Rodale Institute at the time, came to visit and check in on me. When I explained to Jeff the details of my sudden and mysterious health collapse, he told me about Dr. Bill. Though he warned that appointments with him were usually hard to come by.

I didn't have much to lose so I called Dr. Bill's office. By some stroke of luck, there was an opening that Saturday at 8 a.m.

Dr. Bill's office was three miles from my house. He spent two hours with me that day. Turns out, he read the magazine I was running.

I told him my story.

He listened.

He told me fevers were not psychosomatic. My health

collapse had a root cause, and he would help me find it.

And then he said it. Words that would make no sense at the time, but that would once again change the future trajectory of my life.

"We're going to farm your body back to health."

I had questions.

He explained how the body is like a terrain or soil on a farm. Some vector had taken up residence in my terrain. If you strengthen the body, you make the terrain stronger than the vector. The vector can't live. Healthy soil will yield a healthy body.

That's what he would help me do.

This was my introduction to functional medicine.

Dr. Bill saw my body as an interconnected system. He took a whole new approach to my health.

When he learned I'd been prescribed a particular, powerful antibiotic for a full month by my former doctor, he told me that no one should take that particular medicine for more than a few days because longer use could have toxic effects on the body. Right away, he started treating me with protocols for detoxification.

Over the next month, he did more tests, ruling things out one by one.

Eventually, he suggested a test for Lyme disease, one that was not covered by my medical insurance.

Turns out my Lyme numbers were off the charts.

I had been battling Chronic Lyme disease since October. It was now March. I never found a tick bite on my body but later learned that Lyme disease is a very complicated, mysterious, and under-studied illness.

Our mainstream medical community does not fully understand the complexities of this illness, but Dr. Bill had gained a lot of experience treating this illness in his clinic using a Functional Medicine approach.

When I heard the diagnosis, I felt a sense of relief. I wasn't crazy. I finally had an answer.

From there, I started doing my own research around functional medicine, chronic Lyme disease, and anything I could read on these topics. I would bring Dr. Bill information that I found in my own searches and we worked together on an integrated plan to get me on a pathway to healing.

I started feeling better and better, little by little.

I realize now the shift that happened. I became a farmer. I started tending to the terrain of my own body.

I tried everything: hyperbaric oxygen therapy to help reduce inflammation and support healing at the cellular level; infrared sauna to sweat out toxins; ozone therapy to increase oxygenation and fight infection; acupuncture and Qigong to restore energetic balance; intravenous vitamin C to support my immune system; herbs tailored to support my liver and detox pathways; Reiki to calm my nervous system; the Lyme Diet to reduce inflammation and starve the bacteria of sugar and gluten.

I met with a shamanic healer, a yoga teacher, and a psychotherapist.

At one point, I even did something called the Milk Cleanse which is exactly what it sounds like.

I drank nothing but goat's milk for eight days straight. Goat's milk. That one did not work.

But little by little, over time, as I added these new practices, protocols, and teachings into my life, I felt things clearing not just physically, but emotionally.

Old grief, old beliefs, old patterns. Everything softened. My brain fog lifted. My energy returned.

There was not one silver bullet that instantly made me feel better.

An entire collection of practices, protocols, and food as medicine is what got me moving in the right direction.

It wasn't just about killing the Lyme. It was about dying to an old self and being made new again.

None of this happened quickly. It was super gradual, but I was healing.

During this season of my life, I was a volunteer board member at Rodale Institute. I had some exposure to regenerative organic farming.

What I was experiencing in my body felt parallel to the work happening on the farm.

Something new was being born in me. A new frontier for my life was emerging.

I started to understand why I'm here.

As I regained my health, I felt a pull toward something new. It was time to leave my job. I wanted to work on the farm.

As I farmed my body back to health, I wanted to give myself to this.
This work.
This movement.
This way of life.

I wanted others to experience the healing miracle that I experienced.

So I asked for a job at the Rodale Institute.

The answer was a resounding *yes*. They created a job for me and it paid significantly less than what I was making at the time. Like cut my pay in half, then reduce it a little more.

And I said *yes*.

For the first time in my life, money was not the deciding factor. I accepted the job I asked for at the fledgling nonprofit because everything had shifted in my life. For me, there was no going back. I wanted to go forward into a new movement in regenerative organic agriculture.

Because that movement healed me.

So I went to work at my new job. And, little by little, I started to feel better too.

About one year to the day I got sick, on a similar bright, blue, October morning I went surfing for the first time. It had been a lifelong dream of mine to learn how to surf. I made a vow that when I felt better, I would do things that made me feel healthy.

Something about surfing was calling me.

I met my guide Frankie, a former professional surfer, at the north side of the Ventnor Pier. He had a board and a wetsuit for me, and he pushed me into my first wave.

ALIVE.

HEALED.

REBORN.

RENEWED.

REGENERATED.

That's how I felt at that moment.

Sitting in the water next to Frankie, catching those waves, I felt like a new creation.

I'd thought healing was about returning to who I'd been before, but it was actually about being made into an entirely new version of me.

A few months later, my dad and I started building some garden beds in my backyard. Food had become my medicine, so I wanted to grow my own food.

Little did I know that we were using some of the last lumber from the local hardware store that would be available for a while. And a whole new movement was just getting started as a result of the global pandemic.

I spent those days tending to the terrain of my body and the land around me. Growing my own food with my hands in the soil.

Because it turns out my healing—and really, *our* healing— begins in the soil.

JEFF TKACH

*Barn at Rodale Institute,
site of the Regenerative Healthcare Conference*

Chapter 3

THE SOIL IS ALIVE

The soil is the great connector of lives,
the source and destination of all.
It is the healer and restorer and resurrector,
by which disease passes into health,
age into youth, death into life.
Without proper care for it we can have no community,
because without proper care for it we can have no life.

—Wendell Berry
The Unsettling of America: Culture & Agriculture

JEFF TKACH

Just 500 yards to the west of my office, there's a field I walk through nearly every day.

The field spans 12 acres, divided into 72 plots, and for the past 45 years, we've been conducting a research study there. In fact, it's the longest-running study of its kind in North America.[1]

The Farming Systems Trial was established in 1981 by Robert Rodale to address the barriers to adoption of organic agriculture. This long-term study compares organic and conventional farming systems.

This seminal research was largely instrumental in the creation of the National Organic Program housed inside of the US Department of Agriculture (USDA) in 1990. When you shop at the grocery store and see that little "USDA ORGANIC" logo on products, you can thank this study, which gave our federal government the evidence that organic agriculture is a real and viable production system.

When I reach down and pick up a handful of soil from the plots that have been managed using regenerative organic practices, it's like I'm holding pure life.

That tiny amount of earth in my hand contains an entire world.

Soil is one of the most diverse ecosystems on Earth. Over 10 billion living organisms can be found in a single teaspoon of healthy soil—more living things than there are people on the planet.

There are miles of threads, or fungal networks, that weave

everything together, along with billions of bacterial cells (so small that 40 million could fit on the head of a pin),[2] and somewhere between 10,000 to 50,000 other organisms, including protozoa and nematodes.[3]

This soil holds decaying plant and animal matter, essential for nutrient cycling, as well as sand, silt, and clay, which help retain water.

It's packed with macronutrients like nitrogen, phosphorus, and potassium, and enzymes that bind it all together.[4]

This organic soil pulses with life. Through every pore, it breathes, taking in air and water.

If I brought a handful of this soil into our lab and drop it into an aquarium full of water; we'd watch it stay intact. This soil is part of an interconnected web, where everything—bacteria, fungi, and minerals—is held together by a natural glue called glomalin, which gives this soil its structure, strength, and stability.[5] Healthy soil sticks together.

What fascinates me most about holding this soil in my hand is the sense of mystery and wonder it holds. The soil is like an underground universe, alive in ways we're still discovering. Scientists say that 90 percent of soil remains unknown, much like the vast expanse of the universe.[6]

The soil I can hold in my hand—dark, rich, and teeming with life—comes from the plots farmed using regenerative organic methods. Only a small percentage of cropland in America is managed organically, in soil that's truly healthy and alive.[7]

In another part of the field, the soil is managed quite differently, using chemical (or what many call "conventional") methods of farming. In these soils, something entirely different is happening.

For the past 45 years, through our study on 12 acres split into 72 randomized plots, we've identified two very different dynamics happening within the soil of each farming method.

The soil I hold in my hand that is teeming with life is farmed using regenerative organic methods.

Regenerative organic soil

But this is not the reality for most of the agricultural soil on Earth. At least not anymore.

When I reach for a handful of soil from one of the plots that is managed through conventional methods of agriculture, I can't hold it in one shape for very long. It slips right through my fingers.

This soil was farmed with methods that rely on chemicals for fertility and to manage the weeds with very little diversity in the crop rotation. The land gets pushed to its biological limits, stripped of life, and left with no time to regenerate.

This soil falls through your fingers, like sand. It's not alive.

It's clear that this mainstream, conventional way of farming is failing us. In the last 150 years, half of the world's topsoil has been lost.[8] Over 70 percent of U.S. soils are depleted of essential minerals, and the U.S. has lost 57 billion tons of soil to erosion, just in the Midwest alone.[9]

Conventional farming methods also reduce soil diversity.[10] As a result, global agricultural land has lost about 50 percent of its natural fertility.[11] If there is no diversity in our soil, this will lead to the loss of diversity of nutrients in our food leading to the decline of human health.[12]

Scientists predict that by 2050, 95 percent of the Earth's soil will be degraded due to conventional farming practices, with 24 billion tons of fertile soil lost every year—equivalent to four football fields of healthy soil every second.[13]

But when I put my hands in the healthy soil we're cultivating here in the organic plots, I know there's hope for the future. I know we can change things - quickly. Because we're doing it here at Rodale Institute—and it's working.

So, I continue my walk across the farm and head to another field where we're conducting a new study. This one is focused on the nutrient density of the food we grow. Because the soil determines the nutrients in the food, and those nutrients are intended to keep us healthy. For us to thrive and receive the nutrients we need, our food must be nourished by healthy soil. And if the soil is depleted, then we get depleted too. This is why most people today are suffering from what is known as "hidden hunger." Even though we're getting more than enough calories through the standard American diet, we're not getting enough of the vital micronutrients, amino acids, and phytochemicals (the tiny but powerful compounds that we need to truly make us healthy).[14]

The food we eat today is far less nutritious than it was just a few generations ago. The nutritional value of vegetables has declined over the past 50 years due to soil depletion, extractive farming, and the use of synthetic fertilizers.[15]

Modern crops are less nutrient-dense than older varieties. The USDA found that vitamin C levels in fruits and vegetables decreased by 10-20 percent between 1950 and 1999. The study found a six percent decline in protein across the 43 garden crops. It also found a 38 percent decline in riboflavin (vitamin B2).[16]

This is a big problem that's impacting our health. The food from conventional agriculture is making us sick.

Poor diet contributes to 45% of deaths from chronic diseases, like heart disease, stroke, and type 2 diabetes, mainly due to a lack of nutrient-dense foods.[17]

The childhood obesity rate in the U.S. has tripled since 1970, with more than one in five children ages 2 to 19 now classified as obese, also largely linked to poor diet.[18] This sad but true reality is deeply personal to me.

Nearly half of all adults in the U.S. have high blood pressure, a key risk factor for heart disease,[19] with the CDC estimating the total cost of heart disease and stroke at $417.9 billion.[20]

In the last 30 years, cancer rates have increased by 74.3 percent globally, with environmental toxins and dietary factors playing significant roles in the rise of preventable cancers.[21]

Sperm counts have dropped by 50-60% worldwide since the 1970s, with environmental chemicals like pesticides suspected to be key contributors to this decline.[22]

And within two generations, we've witnessed a massive shift in human health. Only 12 percent of the population had a chronic illness in 1960.[23] Today, that number is three in four people.[24]

I believe that this trend is directly related to the food that we eat. And that food is coming from the soil.

The good news?

We don't need complex innovations to solve this trend and move it in a more positive direction. We just need to return our attention to building healthy soil.

This is about two kinds of soil. But it's also about life and death.

One type of soil is alive, the other is dead.

If the soil isn't living, the food won't be either. What we put in the soil is what we get back in our bodies. And right now, what we're getting is sickness.

Your life comes from food, and healthy food begins with

healthy soil. The vitality of the soil becomes your vitality. You and the soil are interconnected—an inseparable unity.

I am more convinced than ever before that our food has an energetic imprint. Life or death. Vitality or sickness. Love or hate.

This connection between soil health and human health landed me an invitation to speak at a healthcare conference for medical doctors in Oakland, California.

I'm not a medical doctor, but I was asked to speak to a room full of 1,200 doctors about how regenerative organic agricultural practices directly impact human health. After my talk, a line stretched from the back of the room to the front.

The doctors had so many questions about farming. I learned that these doctors received only about nine hours of nutrition training in their half-a-million-dollar education.

I told them about how regenerative organic farming practices can restore not only soil health, but the health of our food system and people too. How food can be medicine. How our healing starts in the soil.

I told them my story about healing from a chronic illness by making food my medicine.

To these doctors, this was a radically new concept. They never learned anything in medical school about how better living soil creates better living people.

Driving away from the conference, headed to the airport to fly back to Pennsylvania, it hit me. This was a pretty standard talk that I'd given many times before, yet the interest from these doctors was overwhelming.

Who knew that the very way I had healed myself through food was a revolutionary idea for some of the leading medical minds? There was a hunger for Rodale Institute's science and the use of nutrition as not only a treatment, but for prevention too.

What if, instead of just talking to doctors in a conference room, we brought them to the farm? What if they spent a few days learning firsthand how farming impacts human health?

That question led us to host our own healthcare conference on our farm.

We launched the first healthcare conference of its kind—an accredited medical conference where doctors and healthcare practitioners from seven countries spent four days learning on a farm.

We had a barn full of doctors.

They got their hands in the soil, learned about nutrient density, and explored food as medicine.

They harvested potatoes with their bare hands and then we ate those potatoes for dinner.

Those doctors took the knowledge that they learned on our farm back to their clinics around the world, prescribing food as medicine in a way they never had before.

So now, we regularly host a barn full of doctors and healthcare providers—from all over the world—coming to the farm to learn the most profound lesson of their medical careers: that healing starts in the soil.

These aren't new ideas; they're ancient truths, but in a world that's forgotten them, they feel revolutionary. And as these doctors leave the farm, with new yet ancient wisdom and dirt under their nails, they're not just changing their medical practices. They're bringing the soil's medicine to those who need it most.

These lessons aren't just for doctors or healthcare providers, though. They're for all of us. Soil has a lot to teach us—not just about food and health, but about life itself.

There's a Latin word for soil, *humus*, which shares a root with *humanus*, the word for human. Both trace back to mean earth or ground. In essence, the soil teaches us about what it means to be human.

To understand soil is to understand something deeper about ourselves.

If you connect with the soil, you'll discover it's a teacher, waiting to show you that it's good to be human.

JEFF TKACH

Cover crop at Rodale Institute

Chapter 4

NATURE'S RHYTHM

JEFF TKACH

It all begins with a cover crop.

In January, if you walk around the farm with me, you'll feel the cold wind, maybe see a dusting of snow, and look over the fields, thinking, *Not much going on here.*

The farm may seem still, even dormant. It might look like nothing is happening, but the important things are happening underground, in the soil.

Because in the fall, when we gather the year's last harvest, we also plant seeds for cover crops. Cover crops that usually won't be used for food and have no cash value. Yet, we plant them for winter, knowing they will bring us many benefits.

If you flew a drone over the Rodale Institute in the winter months, our farm sticks out as a vibrant patch of rolling green fields surrounded by a sea of brown, bare soil.

Our farm is emerald green in the dead of winter. Green is a sign of photosynthesis. And the life is coming from the cover crops.

Even in the stillness of the cold season, cover crops are alive and breathing, and the earth breathes with them. There's an invisible exchange happening: carbon dioxide coming in, oxygen going out. Nutrients in, nutrients out. A living, breathing energy transfer is going on between the cover crops, the soil, and the sky.

A cover crop is planted solely for the health and protection of the soil. These crops get planted in fields that would otherwise be left bare between growing seasons. They help replenish and enrich the soil by fixing nitrogen, preventing erosion, and adding organic matter that improves nutrients. In spring, they suppress weeds, reduce pests and disease, improve water retention, and boost biodiversity across the farm.

Even though cover crops don't make any money for the farmer and usually don't produce any food, their value is beyond measure. Like a winter blanket for the fields, they protect and nurse the life in the soil between growing seasons.

The quiet, unseen processes of cover crops set the stage for everything to come. Because come spring, we plant. And by July, the fields are bursting with the productive cash crops that we will take to market.

In January, we nurture and rest the soil with cover crops, so that by July, we reap the benefits in our first harvest. This is why our dominant farming system isn't working.

Every time a farmer grows a cash crop, they extract nutrients from the soil. After harvest, those nutrients need to be returned to the soil. Instead of cover crops, synthetic nitrogen is often used to mimic nature's process and "nourish" the soil.

Synthetic nitrogen that has become ubiquitous in our farming systems was first manufactured during WWII for creating bombs. After the war, these same chemicals were repurposed for farmers to spread nitrogen across their fields.[1] This practice continues today and it's had some unforeseen repercussions.

Like the red tide that we often see in the Gulf of Mexico and Florida. It's intensified by the use of synthetic nitrogen on farms.[2] This chemical finds its way from the Midwest, flowing all the way to the Gulf, creating a harmful algae bloom that causes the water to turn red and produces toxins harmful to marine life, humans, and the environment.

We spend millions and millions of dollars cleaning this up each year. The same could be said about the Chesapeake Bay or any body of water that surrounds our agricultural land in the United States. That's how pervasive and insane the use of synthetic chemicals is.

Cover crops don't do this. In fact, they prevent this. They nourish the soil better than any chemical ever could, along with a million other benefits.

A cover crop has the ability to fix nutrients in the soil which it takes from the atmosphere, improving soil health naturally. Actually, in our research, we've discovered that some cover crops can even do double duty.[3]

Winter wheat, buckwheat, peas, alfalfa, rye, and barley all provide nutrients back to the soil, along with weed control—and they can be harvested too. A true win-win.

Once a cover crop has done its job, the field must be prepared for planting. That's where the roller crimper comes in. Jeff Moyer, our CEO Emeritus and longtime Farm Director at the Rodale Institute, helped bring this innovative tool into widespread use.

The roller crimper gently terminates a mature cover crop by crimping its stems and laying it down as a protective mulch. Instead of leaving soil bare or disturbed, the ground remains covered and biologically active, allowing farmers to plant directly into that residue without synthetic chemicals or intensive tillage. It's proof that innovation in agriculture doesn't require more inputs, just a deeper understanding of how natural systems already function.

Conventional farming using synthetic nitrogen unintentionally leads to water pollution, soil erosion, and loss of essential biodiversity.[4] Regenerative organic farming intentionally leads to healthier soils, reduced erosion, improved biodiversity, enhanced soil fertility, natural pest control, increased carbon sequestration, and better water retention.

The benefits of regenerative organic farming come from following the cycles of nature—and that's what's missing from conventional farming too.

There's a rhythm to how the land, the soil, and the earth naturally thrive and produce while also replenishing.

Plant, cultivate, harvest, rest.
Plant, cultivate, harvest, rest.
Plant, cultivate, harvest, rest.

In conventional farming—and modern life—we find a different rhythm. It often looks like:

Produce, produce, produce, produce.
Extract, extract, extract, extract.

I can see now how I ended up bedridden in January of 2017.

I had achieved what I had thought was the job of my dreams, but I had no metaphorical cover crop in my life. I had no rest. I extracted every ounce of energy from my being, in order to achieve success, whatever that meant.

I didn't know when to stop. I didn't know when enough was enough. I ended up on the couch because I did not have a rhythm to my life. It was all production. All harvest and extraction.

I didn't know how to replenish.
I had no concept that the body needed regeneration.
That the body is like the soil.
The soil is a body.

What I've learned from walking the farm from season to season has changed every part of my life. There's no way to walk as many fields as I have and talk to as many farmers as I have and not be transformed. It's changed the way I see everything—how I live, how I think, and how I show up in the world.

As my body began to heal, the land showed me that food was more than just my medicine. Food wasn't just nourishing me; it was reconnecting me to its source: the soil. I learned to plant, cultivate, harvest, and rest in harmony with nature's rhythms, mirroring the cycles of the land in my own life.

I took email off my phone.
I only look at a screen between 8 a.m. and 6 p.m.
I leave my office at least once per day to walk the land.
I stopped doing meetings on Fridays.

As I've begun to live in this new rhythm, I've discovered that I'm actually far more productive than I used to be. The more I've embraced regenerative practices in my life, the more results I achieved in my work at the Rodale Institute.

In many ways, I do the opposite of what I used to do.

I learned how to breathe.
I literally took classes on how to breathe in a way that regenerates my body.
I learned to get silent at least once per day.
I take time in my day to nourish my body with food.
I spend a half hour every morning preparing my meals for the day.

Sometimes it becomes very clear I need to get some space. So, I go surfing or for a bike ride. I might pack my car and drive two hours to the Jersey Shore, so I can get in the ocean.

This is usually when the greatest challenges in my work sort themselves out. Everything comes back into focus.

When I step away, I get all kinds of clarity I wasn't even asking for.

For many people this is counterintuitive. We've learned that the only way to get ahead is to stay extra hours, work harder, and push through.

It seems backwards to step away. To take a break and return. It might feel or look like you're abandoning your work, but true fidelity to the mission often requires stepping away from it.

When you step away, you take a quantum leap forward.
When you mimic the rhythms of nature, you'll find everything.

If you want a big harvest, plant a cover crop.
If you want the land to produce, give it space to rest.
If you want to move forward, let yourself rest.

The biggest challenges of my work melt away when I take time to sit on my surfboard or ride my bike in the woods. Those challenges solve themselves, revealing their solutions and allowing me to return with fresh perspective and clarity.

When I started incorporating the lesson of cover crops into my daily life, I created energy reserves that I now can draw upon when I need them. I'm working from a full tank, not running on fumes.

Surfing, bike rides, silence, breathing classes. These aren't hobbies. These aren't escapes. These aren't things I sneak away to do. They're mandatory practices. They are my rhythm of life.
They're requirements.

I got sick because I followed conventional wisdom, which tells you to push yourself, override your needs, and neglect self-care.

For so many of us, burnout, stress, and anxiety are just how the world works. Lots of actions and behaviors get excused. Like when I was younger, pushing myself so hard, trying to climb the corporate ladder and give it my all. I thought that's what I had to do.

But that's not true.
How it works is cover crops.
How it works is rest.
How it works is rhythm.
How it works is living regeneratively.

The delusion that we can endlessly extract from ourselves or the earth is what's killing us in a million different ways.

The natural world follows a different rhythm.
A regenerative one.

If we want to live in a world free of human and planetary degradation, we need to understand the rhythms of regeneration. We need to follow the cycle of plant, cultivate, harvest, rest—on the farm and in our own lives.

When you're living in proper alignment with the Earth, your life becomes a rhythm, which includes your job. The boundaries between personal and professional dissolve. Not because you're working all the time, but because you have proper balance.

These aren't options.
These aren't alternatives.
This is how the natural world actually works.

This is the invitation for all of us.
To live regeneratively.
To move and act differently.
Do you want to thrive?
Follow the rhythms of nature.
There, you'll find everything.

As you begin to understand how regeneration works, you're taken into a whole other world.

After rest, comes a return.

The land needs rest. It also needs nutrition. Otherwise known as this beautiful thing called compost.

Compost pile at Rodale Institute

Chapter 5

STEAMING PILES OF LIFE

the one world
we all belong to

where everything
sooner or later
is a part of everything else

—Mary Oliver
"Poem of the One World"

It's July 1st, 2017, and today is the first day of my new job at Rodale Institute.

I find myself riding shotgun in the cab of a John Deere 6195R tractor as it pulls a Sittler 1014 compost turner. I'm sitting next to Rick Carr, our master composter and a true expert in compost science, production, and utilization.

As I begin my new life at Rodale Institute, I make it a point to spend time with every department and function across the organization. I make myself a student of regenerative organic agriculture.

While riding with Rick, I learned that local municipalities drop off yard waste to our farm bringing leaves, grasses, branches, and other backyard material collected curbside that would otherwise end up in landfills.

Every autumn, dump truck after dump truck rolls onto our farm, and Rick and the compost research team get to work blending it all together into a cacophony of black gold.

All that compost is made into piles called "windrows" where they cook up a living recipe using ingredients that most people discard as waste. As a society, we waste, waste.

On their own, an orange peel, a handful of leaves, and a scoop of animal manure don't do much. But when you mix them all together and start churning, turning, and blending - something magical starts to happen.

Everything in the pile begins interacting with everything else, oxidizing and breaking down.

The mix of carbon-rich "browns" (like leaves and straw) and nitrogen-rich "greens" (like food scraps and manure), combined with the right amounts of water and oxygen, creates the perfect conditions for life to take over.[1]

Every couple of weeks, the team drives the compost turner through the piles at a screaming 0.1 miles per hour. As the blades churn the material, the piles begin to release steam.

Microbes get to work, and the pile generates heat. Sometimes up to 170 degrees, even in the middle of the winter.

In January, when I pull onto the farm, I can see the piles steaming in the morning. Proof that even on a freezing cold day, they're cooking away. Rick told me that you can toss an entire dead animal—a deer, a pig—into these piles, and within a few weeks they're fully cooked down into compost that you can spread onto your garden.

In these piles, all the discarded scraps have come back to life as compost. A crumbly, sweet-smelling, nutrient-rich compost.

There is very little waste on our farm. Instead, we turn it into food for the farm. Farms need to be fed.

Conventional farming doesn't do it this way.

Conventional farming depends on synthetic fertilizers, which are nutrients made in a laboratory, extracted from petroleum and other fossil fuels that are mined and shipped across the globe to apply to farmland to grow corn and soybeans for ethanol to put in the gas tanks of our cars.[2]

Are you with me?

These synthetic fertilizers attempt to feed the farm by targeting only one plant at a time. The process for making these synthetic fertilizers is highly extractive and can be degrading to the environment.

Inevitably, the fossil fuel-based, synthetic nutrients can find their way into streams, rivers, lakes, and oceans.

Excess nitrogen and phosphorus from agricultural runoff have been shown to accelerate algae growth beyond what aquatic ecosystems can handle, depleting the oxygen that fish and marine life need to survive. This process contributes to what scientists refer to as "dead zones" in oceans and waterways, which have become increasingly more widespread.

One of the largest documented dead zones is in the Gulf of Mexico, driven in part by runoff from farms in the Midwest, and has spanned more than 4,000 square miles in recent years.[3]

This runoff can affect drinking water sources.

In addition to nutrient pollution, certain agrichemicals have raised environmental and public health concerns. Atrazine, a widely used herbicide in conventional farming, has been the subject of ongoing scientific debate regarding its potential endocrine-disrupting effects. The European Union banned atrazine in 2004 due to concerns about groundwater contamination, while it remains in use in the United States.[4]

This is the cost of a system built on extraction and disconnection—from the land, from each other, and from the natural cycles that sustain life.

Conventional farming is just one example. But it reflects a broader cultural mindset.

Our modern culture is built on extraction.

Take, take, take.

Waste has become an afterthought.
Waste has become an inconvenience.
We have become divorced from our waste.
Landfills are often located where no one can see them.

And we don't understand the consequences of our actions.

Humanity is using resources 1.7 times faster than the Earth can replace them.[5]

The average American throws away approximately 4.9 pounds of trash every day.[6]

Every day, approximately one billion meals are wasted globally.[7]

51.4 percent of landfill waste could instead be composted and returned to the earth.[8]

This type of extraction becomes a way of life for many. It does not bring peace, fulfillment, or satisfaction.

On the deepest spiritual level, endless consumption followed by thoughtless waste makes us miserable.

But finding your way out of this is, in many ways, is *the question* of the modern world.

THE FARM IS HERE

The farm tells us that a different life is possible.

On the farm, what I notice—season after season—is that all things are constantly being made new. On our farm, there is no waste and there is no death.

For me, composting serves as a daily reminder that death is not an end point, but a return.

Composting is a mysterious, mystical, and fascinating process. Where waste becomes life.

Gradually, we lose the sense that there even *is* waste. Because everything belongs.

Our waste becomes the solution.
What we discard becomes new life.

Every day, I get to experience new creation through my participation in regenerative organic agriculture.

During my season of illness—when I wasn't sure if I was going to make it—the fear of death felt present. But as I began to heal my body through food and a deeper connection to the earth, something in me shifted.

The fear started to fade.

Being so close to the cycle of life unfolding on the farm each day became its own kind of medicine. It reminded me that death is not an ending, but a transformation.

As I became more and more tuned into these cycles of participation, it completely transformed my life and my relationship with death.

On the farm, we're invited to participate in the rhythm of life.

Nothing is wasted.
Everything has a place in the cycle of life and we all want to participate in that cycle.
It's where the joy, the peace, and the life is.

Participation makes us more at home on the Earth, in our bodies, and in the sacred rhythm of return.

Being a part of these cycles makes us less violent, more connected to the natural world, and more attuned to the sanctity of all life.

Nature feeds us, so we feed it.
So, how do you move from extraction to participation?
How do you move from taking to contributing?

Start with your food.
Start with your body.
Start with compost.

Start with you.
Put a bowl on your counter.
Start collecting your food scraps.
Put them in the compost bin.

And witness how this simple act transforms your life.

*Dario Fornara, Research Director,
European Regenerative Organic Center*

Chapter 6

THE GUTS OF IT

Fertility of the soil is the future of civilization.

—Sir Albert Howard

A planet with two suns was discovered.

Yes, astronomers have confirmed the discovery of a planet orbiting two suns. This planet, named Kepler-16b, is about 200 light-years away and circles a pair of binary stars.[1]

It sounds like something out of a science fiction novel or a *Jetsons* episode. Something hard to imagine or comprehend in our lifetime.

While exploring the mysteries of the cosmos is mindblowingly fascinating, what captivates me most is the mystery unfolding right here on Earth. In a world obsessed with space travel, algorithms, and A.I., we're overlooking the biggest mystery of all: the one unfolding right beneath our feet.

We want to discover signs of life out *there*, but a greater mystery is unfolding right *here*.

We live on a planet rife with ancient, complex life forms that we're only just beginning to understand. Our planet holds completely undiscovered worlds.

For most of modern history, people have seen the soil as a means to an end. A stagnant, passive foundation to grow plants. But what we've helped discover is that there's actually an entire undiscovered world of life, right beneath our feet.

Soil is home to 59 percent of all plant and animal species on Earth.[2] It supports a wider range of life than any ocean, populated with 60 billion living beings. Most of them cannot be seen by the human eye.

Just like the discovery of the world above us, this revelation of the world below us changes everything we thought we knew.

We depend on this complex network of life for our survival. 98.8 percent of the calories we need to survive come from the soil.[3] Of the 29 elements that are essential for human life, 18 come from the soil.[4] Without soil, there's no life. This thin, delicate layer of Earth is what's keeping us alive. It's where life comes from.

There's pioneering research being done at the intersection of the soil and human microbiome level. The life in the soil comes from its microbiome and so does ours. In fact, plant roots resemble our guts.[5]

In the soil, plant roots are surrounded by a community of microorganisms, like bacteria, fungi, and worms, living around and within plant roots. They help the plant grow, absorb nutrients, and protect the plant from disease.

In our bodies, our guts contain trillions of microbes, like bacteria, fungi, and other tiny organisms, living in the gut and on the body. They help us digest food, absorb nutrients, and protect us from disease.

Plant roots and our guts function in similar ways. They both need the right amount of water, oxygen, and pH balance to survive. They both need the same kind of bacteria, like Firmicutes and Bacteroidetes.[6]

Dr. Dario Fornara, our research director at the Davines Group-Rodale Institute European Regenerative Organic Center (EROC) in Parma, Italy, explains, "Soils are major hubs of biodiversity, and there's growing evidence that soil, plant roots, and the human gut microbiome are connected in similar ways."

It's not surprising that plants have a big effect on our guts.

Some studies suggest that the plants we eat can actually shape the microbes in our gut, especially when we eat them raw or fresh from the soil. In fact, up to 2 million bacteria can be found on a single gram of fresh produce.[7] And for centuries, human waste, also called "night soil," was used as natural fertilizer, recycling nutrients like nitrogen and phosphorus back into the soil, helping to feed the soil microbes.

In essence, the soil feeds us and we feed the soil. Human health is scientifically dependent on soil health.

In our modern world, we've lost sight of this. We've only been talking about the soil being alive for about the past 70 years. And we've only started to use the words "soil health" in recent years.

For thousands of years before conventional agriculture took over, indigenous people produced food in a way that centered on natural cycles. They stewarded healthy ecosystems and communities and recognized all life, including the life of the soil, as sacred.

We can improve the health of our soil very quickly. And when we do, we'll improve our health too.

But first, we have to face our current reality.

For the past half a century, a massive, industrial economic system has ravaged the soil.

In the 1970s, Earl Butz told farmers to "get big or get out." He was appointed Secretary of Agriculture in 1971 by President Richard Nixon. He eliminated a federal program that incentivized farmers *not* to plant all their land and to set some farmland aside to promote biodiversity.

Instead, he encouraged farmers to plant every last inch with cash crops like corn. He decided the most important metric of success on a farm was *yield*. Yield refers to how much product a farmer could produce on their acreage. His policy shifts destroyed so much life on the farm. And he paved the way for growing food to become a commodity business built on extractive principles.

From there, a few corporations took advantage of these new dynamics in agriculture and built proprietary business models around them. They invented and patented chemicals to grow food faster and cheaper. A growing body of peer-reviewed research has examined potential ecological and biological impacts of chronic herbicide use. Studies have explored potential effects on soil biology, water systems, and microbial communities. Research synthesized by Rodale Institute highlights concerns about the persistence of certain chemical herbicides in soil and water, as well as their potential influence on ecosystem health and microbial balance.[8]

These findings are a part of ongoing scientific conversation about the long-term environmental public health implications of industrial agricultural inputs.

As a society, we've become so accustomed to cheap food that we now expect it. We've lost sight of the value of nourishing food.

The truth is, nothing drains your health—or your wallet—like eating nutrient-poor food, grown in nutrient-poor soil.

When we destroy soil health, we destroy human health too.

And the costs are immeasurable.

Approximately 6 in 10 young, 8 in 10 midlife, and 9 in 10 older US adults report one or more chronic conditions.[9]

We're seeing spikes in rates of dementia, depression, and substance abuse.[10] Some studies have found links between the use of herbicides in conventional farming and increased cancer rates.[11]

We wonder why there is so much hatred in our society.
We wonder why we can't agree on anything.
We wonder why people treat each other so poorly.

It's because we're sick.

It's because the food we're being sold doesn't promote life. The food we're being sold promotes death.

I believe that food has an energetic imprint. Food holds energy. Energy that either promotes love or hatred, regeneration or degeneration, health or sickness, life or death.

Many of us are suffering from hidden hunger. We're eating more than enough calories, but there's a lack of essential nutrients in our food such as iron, zinc, vitamin A, and B vitamins. It's possible to have a full stomach but still be dying on the inside.

If you don't feel good, then it's hard to get along with others.

We need to know where our food is coming from, how it was produced, and what nutrients it contains.

This is not the fault of farmers.

Farmers do not wake up each day with the intention of making people sick.

This is the by-product of the failure of an entire system. When Earl Butz made those changes in federal policy, he made it nearly impossible for diversified family farms to continue to exist financially.

Our food system has created significant unintended consequences.

This is the reality we're facing as a society and as a result of an industrial food system that is predicated upon extraction.

We know from the science at Rodale Institute that diversity on a farm promotes health.

That diversity is what I experience on my daily farm walks. There's forested areas on our farm, fields we let grow into a meadow to attract birds and beneficial insects, and a very diverse rotation of crops being planted throughout the year. We've created a rich and diverse ecosystem in which life can thrive. Just like the soil and just like our bodies.

If we want to be healthy, we need diversity.
If we want a healthy planet, we need life on the farm.

If we want life in us, we need life in our food.

It's all connected.

Life begets life.

So, how do we build life on the farm?

It's simple.

Crop rotations.

Crop rotations are the practice of planting different crops sequentially on the same plot of land. For instance, if a farmer has planted a field of corn, he might plant soybeans next, since corn consumes a lot of nitrogen while soybeans return nitrogen to the soil.[12]

Most farmland in the US is managed through the following rotation:

CORN CORN CORN SOY
CORN CORN CORN SOY
CORN CORN CORN SOY

If a farmer plants the exact same crop in the same place every year, as is common in conventional farming, the same nutrients are continually drawn out of the soil. Pests and diseases happily make themselves a permanent home as their preferred food source is guaranteed.

With farming like this, more and more chemical fertilizers and pesticides are needed to keep yields high while keeping bugs and disease at bay. This way of farming kills the soil, leading to habitat destruction, pollution, and harmful unintended consequences.

At Rodale Institute, we use the following rotation:

HAY RYE CORN SOY OATS WHEAT
HAY RYE CORN SOY OATS WHEAT
HAY RYE CORN SOY OATS WHEAT

The way we plant is prescriptive. We prescribe plants for each field, depending on what nutrients it needs in that part of the cycle.

This type of crop rotation helps return nutrients to the soil without synthetic inputs. It also works to keep pests and diseases away too.

We prescribe all kinds of rotations. A simple rotation might involve two or three crops, and complex rotations might incorporate a dozen or more. Our 45 years of continuous science has proven that the more complex a crop rotation, the healthier the soil.

Life in the soil thrives on variety.

Now, even in the organic world, new ways to commodify are popping up. We are constantly getting requests to ask us to try out, test, and review products, like soil amendments and magical potions bottled up under the moon.

Our answer is no. Regenerative organic agriculture requires three things: cover crops, compost, and crop rotations. Everything a farmer needs already exists on the farm.

It's about building the system on the farm.
A systems-based approach to agriculture.
There's no need for any magic potion.

Food carries energy into a human.

When food is grown in living, healthy soil, your food is infused with vitality, nutrients, and life-supporting compounds.

When grown in depleted or chemically treated soil, your food may look the same but lacks the vibrancy and depth that nourishes us on every level—cellular and energetic.

A new initiative at Geisinger provided high-risk, chronically ill patients with organic food for six months.[13] A pilot study found that healthcare costs dropped by 80 percent for participants when they made the switch to healthier, more nutrient dense food.[14]

Soil health, human health, the farm, and you—it's all one system. When my doctor told me I had to farm my body back to health, I quickly came to learn what he meant because I came to see that everything is connected to everything else. The body is a complex ecosystem.

We have the ability to promote life in our bodies.
We have the ability to promote life on farms.
We have the ability to change how farmers produce food because we as consumers are the ones demanding the food itself based on how that food was produced.

So many people believe the world is getting more and more complex.

In some ways, yes. But in other ways, we are being invited to live a simpler, healthier, and more connected life.

Where is your food coming from?

How was your food produced?
These are very basic questions that we all need to be asking.

These questions can change the fabric of our society very quickly.

But we need to start asking them.
Your guts are like plant roots.

What digests life in you is linked to what grows life in the soil.

You and the soil are made of the same stuff.
Your guts are one and the same.

Healing the soil is healing ourselves.

Ramon Madrid, Visual Content Creator, Rodale Institute

Chapter 7

A BOX OF VEGETABLES?

If fungi were to speak…they would tell us what they show us, which is that really the death of an organism is the beginning of countless others; that there is no end to life, just a constantly shifting substrate.

—Giuliana Furcias quoted in *Is a River Alive?* by Robert Mcfarlane

JEFF TKACH

People have so many questions for me about this regenerative organic farming movement.

I love this movement, I want to be a part of it, but how do I actually do this?

How do we feed people organic food when it is so expensive? How do we tend to our bodies when we have bills to pay? How do we care for the Earth in a technology-centric world?

How will organic agriculture produce enough food to feed a growing global population?

How do I live a more regenerative life?

I hear them and I hear you.

But I believe that there is one question that is the question behind all of the other questions...

How much of the world's ache might actually be traced back to the soil?

I know the challenges of the modern world are real. Maybe when you look at your budget, you feel endless pressure to earn. You desire to live simpler and more connected to the Earth but the demands of your job, your family, and your life are constant and make it so difficult for you to slow down and connect.

The farm tells us a different story. A totally new way of looking at time, budgets, and resources.

A story that most of us were taught was not possible.

Let me show you how this works.

Ramon served in the US Army for 11 years. After growing up in New Orleans, with a large family and a single mom, he was searching for purpose. He chose to be of service, but when he left active duty (which included tours in Iraq and Hurricane Katrina), he hit rock bottom, battling PTSD, depression, anxiety, insomnia, and alcoholism.

This guy went to war.

Through his own personal healing journey, Ramon discovered the Rodale Institute's Farmer Training (RIFT) program. Reluctantly, he decided to apply for the nine-month program wondering: *Will this even work?*

Meanwhile, here at Rodale Institute, our students were growing all of this nourishing, nutrient-dense food. We found ourselves asking: *What do we do with all of this excess produce that is a direct result of our educational program?*

At the same time, Jackie and I felt a personal call. Even though my job centers around regenerative organic agriculture, we'd talk about how there were so many people out there who could benefit from better food: *How do we feed those who are most in need of this nourishing food right in our community?*

On his first day of the program landing in Pennsylvania from New Orleans, Ramon was nervous. He quickly found community amongst his fellow veterans who were also learning how to farm.

Our Veteran Farmer Training Program enables people who served in our armed forces to gain a free education, housing, and a living stipend. This program was designed to teach veterans skills in organic agriculture and place them into jobs on farms upon graduation.

We have a program that heals people who have been deeply traumatized by war.

How do soldiers heal from PTSD?

In my experience, healing begins by getting their hands in the soil.

After nine months of being out in nature, working the land, and learning about regeneration, Ramon felt better.

So much so, that he decided to return the following year to give back. Through our fellowship tract, a second year of the RIFT program, Ramon dedicated his time to teaching and supporting other new students in the program.

It gets even better. After exploring all of the facets of Rodale Institute's mission, Ramon was given the opportunity to combine his passion for organic agriculture with his other passions for photography and videography. Now he shares the story of regeneration through visual communications. Think drone footage, educational content, and high-impact storytelling.

Ramon came to Rodale Institute seeking healing. The farm healed him, and now, he helps heal the farm. All the knowledge, support, and community that he received, he's giving it all back by inspiring others worldwide to reconnect with the land and consider farming as a pathway to healing, transformation, and purpose.

It's noon on Thursday.

As I do on most days, I'm out walking the fields in between meetings.

When I reach the back of the barn, I see a party erupting.

Rodale employees are playing music, sharing laughter, and forming a line like clockwork for the weekly vegetable distribution.

This all started with a simple idea: What if we put the extra vegetables our students were growing to good use?

Why not feed the people who are working on the mission? CSA, short for community supported agriculture, is a partnership between consumers and farmers where consumers purchase a "share" of a farm's harvest in advance, providing the farmer with financial support to plant crops and operate the farm. In return, consumers receive regular deliveries of fresh, seasonal produce, often in a box or bag, throughout the growing season.

The only difference is that our CSA does not have a monetary cost to the employees who participate. It is a system based on reciprocity.

Every employee at Rodale Institute gets nourished. Students are feeding us and we feed them. Our students get a free education, and all the food grown is given to Rodale employees and their families—for free.

What began as a CSA offering to employees and students has now taken on a life of its own. What was once an idea has become a community.

The farm has shown me there's always something new and fresh that leaders—and everybody else—can do with the resources around them. And that there are always resources to draw upon beyond what is listed on a balance sheet.

Jackie and I love to meet local organic farmers. Each year, we join a CSA outside of Rodale Institute to support organic agriculture in our wider community.

Every farmer we meet knows someone who could use a share of vegetables but can't afford it. So, we buy a share and we give a share away, paying it forward anonymously.

Imagine if everyone with the means chose to do this. What kind of world could we create if everyone who could help, did? At a time when many people are struggling to pay for healthcare, why don't we provide true healthcare with organic, nutrient-rich food for those who need it?

The farm has shown me that the arguments we use to avoid change don't have any merit. Things like:

Not everyone can afford to buy organic food.
Organic agriculture cannot feed the world.
We need chemical agriculture in order to meet the demands of our ever-growing population.

The farm crushes the misconceptions of capitalism.
The farm constantly shows us that there are endless resources that are possible and available to us.

The farm is the ultimate display of reciprocity.

How do we help socially marginalized people afford organic food?
How do we nourish our neighbor?
How do we feed our world?

We share.
We give, then we receive.
Then we repeat.
Over and over again.

When I was sick, I had to receive.
Once I healed, I was healed to give.

When we feed nature, nature feeds us.
When we feed each other, we create community.
When we create community, we build the foundation for a better world.

A new economy is being birthed. One that is less about extraction and more about reciprocity.

It's about all of us participating in an endless exchange of giving and receiving.

This concept is actually quite ancient and proven. For centuries, people lived in systems based on mutual exchange and community care. They shared resources and supported one another, long before the invention of money, before the invention of the plow, and before factories became the norm.

This ancient way of living honored balance and reciprocity as the foundation of community and survival. I believe we're heading into a new way of living that embraces these ancient truths.

As the late Pope Francis put it, "Let's say no to an economy of exclusion and inequality, where money rules instead of serving. This economy kills! This economy excludes. The economy destroys Mother Earth."[1]

His words are a powerful indictment of the systems that have commodified life itself and they serve as a call to return to an economy rooted in care, justice, and reciprocity.

The farm shows me this new economy every day.

In 2008, Ecuador became the first nation to recognize the rights of nature in its constitution, with specific articles granting nature the right to exist, persist, and be respected.[2] Since then, New Zealand, Panama, India, and Bolivia have passed similar laws acknowledging the intrinsic rights of ecosystems.[3] This means that nature, including rivers, forests, mountains, and the air, are no longer seen solely as property, but as entities with the legal rights to exist, flourish, and evolve. These rights are enforceable by all individuals, communities, peoples, and nations, empowering people to speak on behalf of nature in courts and ensure its protection.

This makes sense because as the farm teaches us, nature is alive and science is starting to prove it. For example, a recent study found that mushrooms speak to each other using up to 50 word-like electrical impulses.[4] Scientists have discovered that trees communicate through an underground network—often called the "Wood Wide Web"—sharing nutrients and warnings, using chemical signals and electrical impulses.[5] And studies show that plants can "hear" the sound of water and sense the vibrations of a caterpillar chewing their leaves, triggering defense mechanisms like releasing chemicals or changing their growth.[6]

Seeing how deeply sentient nature truly is, it's clear that our relationship with the natural world must become the foundation of how we build and measure our economy going forward.

It's time for our nation's farming system to become the baseline for our economy.

It's time for the health of our soil to be prioritized.

It's time to pay an ecosystem service tax on everything that we purchase.

Our current economy focuses on *How much can I buy?* Instead, we can shift to asking *How much can I give and receive?*

The farm shows us how to take a portion of what we've been given and give it to someone else. If we look closely enough, all the resources we need are around us all the time.

Truthfully, most of us are already doing this in some way. Americans have given seven times more to others in recent years than they did 60 years ago.[7] In fact, $592.50 billion went to missions we care about in 2024.[8]

Researchers have shown that people feel happier when they spend money on others versus spending on themselves.[9]

But generosity and reciprocity aren't just about money. We can give our time, our skills, our attention, our extra vegetables, and anything we have in abundance. When we offer what we have, we create connection, impact, and purpose in ways that money alone never could.

I see this every day at Rodale Institute. There's always new and fresh and innovative things we can do that don't require more money or a bigger budget. We have unlimited resources if we just look a little closer to what's hiding in every corner of our world. Hiding behind every corner is a resource that could be freed up that does not cost anything but that gives us everything.

Buy a CSA share, give a CSA share.
Eat your veggies, compost your scraps.
Learn a skill, teach someone else.
Borrow a book, pass one along.
Eat a home-cooked meal, prepare one for someone else.

The list goes on and on and on.
The possibilities are endless.

This new economy declares that we already possess vast resources that we simply need to free up to be given and received.

There's always, always something we can do. Something you can do.

You don't need more. You just need to see differently, give freely, and trust that what you offer always comes back around.

It's how the farm works, an endless dance of giving and receiving.

Maria Rodale, Author and Co-Chair,
Board of Directors, Rodale Institute

Chapter 8

THE TRUTH ABOUT REGENERATIVE

When we demand organic, we are demanding poison-free food. We are demanding clean air. We are demanding pure, fresh water. We are demanding soil that is free to do its job and seeds that are free of toxins. We are demanding that our children be protected from harm.

—Maria Rodale, *Organic Manifesto*

Maria Rodale was born onto America's first intentionally organic farm. Her grandfather championed the term "organic" as we know it today. And her father developed the term "regenerative" that is now being co-opted and greenwashed by some of the largest food companies in the world.

We need to understand the Rodale legacy in order to fully grasp the origins of regenerative organic agriculture and the moment of inflection that we are experiencing within the food industry.

I sat down with Maria to learn how her family birthed a movement that is transforming agriculture as we know it:

"I grew up on the farm where my grandfather founded the modern organic farming movement. Some of my earliest memories are of baby pigs in the barn, shelling freshly grown peas, and eating my absolute favorite vegetable, lettuce.

As a kid, I loved salad so much that I used to just shove it in my mouth with my fingers. Still do.

It was the 1960s, so in the summer we were sent outside to play in the morning and didn't come home until my mom blew the car horn for dinner. For real. She did that every evening.

I was a wild, free-range kid, completely unaware at the time of how unique our life was. Everything around me was farm fresh and organic. It was a way of life I didn't realize was extraordinary until much later.

Things changed when I started school. Kids made fun of me

because of my grandfather's reputation. At that time, he was being called a "quack," "unpatriotic," and a "manure pile worshipper." Other kids thought that if they came to my house, they'd be forced to eat tofu. I didn't even know what tofu was!

That was the first time I noticed a cultural disconnect.

I was living on this incredible farm, but the outside world mocked us. My family was seen as weird, even ostracized. To top it off, my grandfather was Jewish and my grandmother was from a poor coal-mining town, although they both became sophisticated New Yorkers, so they looked nothing like the conventional farmer stereotype of the time. They weren't like that at all. They weren't hippies, either. But they were bohemian.

And yet, where we are today, it's clear: my family knew the truth before the world caught up. Organic farming was always the way forward.

When my grandfather, J.I. Rodale, passed away in 1971, I was nine and two significant things happened.

First, my father, Robert Rodale, became determined to validate through science what his father had been espousing through anecdotal evidence. My grandfather had been ridiculed so much in his lifetime. My dad wanted science to back up the organic movement, so no one could dismiss or mock us anymore.

At the same time, my parents realized we needed a new farm. Our original farm had become like a fishbowl in our community. So many people were dropping by the farm, (sometimes even looking into our windows during dinner!), so that they could pay homage to and tour the farm that was birthing the hippie health food movement.

And besides that, our little farm wasn't big enough for the kind of scientific study my father envisioned. That's when they bought what we first called the "New Farm." It eventually became the Rodale Research Center, and in 1981, the Farming Systems Trial officially began.

I interned on the farm as a teen. I was doing things like harvesting corn samples, drying them, and grinding them up for analysis, so the scientists could run their tests. Dr. Dick Harwood was there that summer, already laying the groundwork for the Farming Systems Trial, which he would go on to lead as the chief scientist.

Maria Rodale as an intern at Rodale Institute, 1979

After that, I became a single parent and went to college. When I graduated, I moved to Washington, DC to work for Fenton Communications, a politically progressive PR company. Back then, I wasn't seen as any kind of heir to the Rodale family business. I was the troublemaker in the family.

But the day I took my last college exam, my brother, who was seen as the heir apparent, died of AIDS. Suddenly, my dad turned to me and said, "I'm counting on you." But I already had my job lined up in D.C., an apartment, and daycare for my daughter. I insisted on going to Washington anyway.

This was when I first started getting more involved with the Rodale Institute (in a family business, one is never not involved). My father would come to D.C. to meet with people at the USDA and for Rodale Institute board meetings, and sometimes I would be a part of those meetings. Or we would get together while he was in town, eating at the restaurant Nora, which was a famous organic restaurant. But he gave me a deadline: "You have one year. Then you need to come back."

So I did. I was given the title: Coordinator of Regenerative Spirituality and Psychology. At the time, my personal interests leaned more toward psychology and spirituality than farming.

People always assume that because I was born into the family business, and talks about soil health at the dinner table, that I knew all there was to know about organic. I considered myself more of an artist and enjoyed gardening as an art form. Although it was in the air and in my DNA, it didn't fully become personal for me until later. When it stopped being just my legacy and started being my *calling*.

I call it my *Bisquick moment*.

I was 20, a single mom, just starting to learn how to cook. I had a baby to feed, and one night I picked up a box of Bisquick, thinking I'd make chicken and biscuits (just like the way my mother did, by the way). And I looked at the ingredient list and it just hit me. There's nothing organic in this food. Nothing nourishing. Nothing I'd want to put in my body or my daughter's body. That was the moment I decided:

I'm going to learn how to make organic biscuits from scratch. I'm going to grow vegetables. I'm going to feed my child the way I wish the whole world would eat. Delicious, organic, real food.

That's when the mission first became mine, not just something I'd inherited.

A few years later, Fenton Communications worked with the Natural Resources Defense Council (NRDC) and Environmental Working Group on what became widely known as the *Alar scare*. Alar, the trade name for daminozide, was a chemical used in apple orchards, and research showed it was potentially harmful, especially to children. In 1989, NRDC released a report raising concerns about potential cancer risks to children from dietary exposure to Alar-treated apples.[1] Apple juice, applesauce—the staples of every kid's lunch box—were laced with this stuff. It was a national scandal. I worked on the advisory board with NRDC's Mothers & Others for a Livable Planet, alongside Meryl Streep, who publicly advocated for stronger pesticide protections for children and spoke before Congress on the issue.[2] The controversy became a defining moment in the national conversation on food safety.

At that point, I said to both my father and David Fenton, "We can't just expose this. We have to support the farmers who are going to be hit hardest by this news. We need to help them transition."

So I teamed up with George DeVault, the editor of *New Farm* magazine, and we launched an 800 number—a real lifeline. Farmers could call in and get support, advice, and coaching

on how to move away from chemicals like Alar and start exploring organic methods.

For me, this was a turning point. It was the moment I saw that we weren't just talking about soil or food. We were talking about regenerating *systems*. About *people*. About *health* and *justice* and *change*.

That's when I really knew: This wasn't just a family business. This was *my* life's work.

I had grown up on this beautiful, abundant farm, surrounded by delicious food and vibrant life. But the way organic was perceived in the broader culture was completely different. It was associated with ugly produce and bland food. No fat, no salt, no sugar. The idea of "healthy food" felt joyless.

So my pushback was this: I wanted to show and prove that organic could be delicious, beautiful, sensual, stylish, and even, luxurious.

I wanted organic to appeal to normal people living normal lives, not just ascetics on the fringe. Regenerative organic agriculture could be vibrant, joyful, and full of pleasure. A way of life. A healthy and happy way of life.

I think what I brought to the movement was balancing all the masculine energy with feminine energy. While my grandfather and father did the visionary thinking, it was my grandmother and mother that did the gardening, cooked the food, and made everything possible. I wanted to do both.

Back then, I didn't have any real agricultural experience, but I had a garden. I had a kid. And I was a working woman. At that point, in the 80s, my dad was running both the publishing company and the Institute with the help of John Haberern.

The *Regeneration Project* was my father's bridge between the two organizations.

The *Regeneration Project* was born around the time of the U.S. farm crisis of the 1980s. Farms were going under, families were devastated, and there was a spike in farmer suicides. High interest rates, plummeting commodity prices, and rising debt made survival almost impossible for many farmers, especially in the Midwest.

My father was already thinking beyond farming. He was seeing how the land was healing through the results of the Farming Systems Trial, and it clicked for him that regeneration wasn't just about soil—regeneration began on farms but seeped out into communities, into people's minds and hearts.

At Rodale Institute, everything was connected: soil health, human health, community and personal well-being. It was all one ecosystem. He used to say, *"If you get a cut on your arm, your body knows how to regenerate. It scabs over, maybe leaves a scar, but it heals. That's what nature does, it heals."* That's how he defined regeneration, as this innate, natural capacity for renewal and healing. Most of the people in the publishing business thought he was crazy. Even though the business made the most money from health magazines and books, they didn't get it. But he did. He believed in it so deeply he started the *Regenerative Management Institute* and held retreats in the Poconos where people brainstormed about applying regeneration to everything from leadership to business to spirituality.

They also developed a training program for all Rodale employees based on these concepts. It went way beyond agriculture. For him, regeneration was his spiritual path.

One of the key ideas my dad introduced was the *capacity analysis*. Instead of listing all the problems and what was missing, the approach was to inventory what the community *already had* and build from those strengths. It was a hopeful, regenerative way of looking at things, rather than a deficit-focused or extractive mindset.

During this time, my dad did tons of international traveling, meeting people from all over the world. He became very contemplative, especially after my brother died in 1985. His spirituality was deeply rooted in a mystical understanding that everything is connected.

Even though he was the head of the publishing company, that wasn't really his true love. His true love was spreading the word about regenerative organic agriculture, traveling, meeting really interesting people, and having great conversations.

His mother came from Lithuania and Russia, and his grandfather was from Poland, so he very much felt like that part of the world was a kind of homeland for him. Back when Russia was still the USSR, he wanted to help the Russian friends he'd made while on the 1968 US Olympic Skeet Shooting Team, who were struggling and needed access to more food. He decided to launch *New Farm* magazine in Russia. But because of communism, he couldn't just bring in his own money and start a business. Instead, he and his Russian friends came up with this wild idea to start a sausage factory that would raise money to fund the magazine. The magazine was a vehicle to spread the message of regenerative organic farming to farmers in Russia that were struggling to survive.

On his last trip in 1990, he was in a van with the driver, his key business partner, and a translator who was also a partner, on the way to the airport to fly back to the United States. On the highway in Moscow, a military truck pulled out on the opposite side of the road, and a bus swerved into oncoming traffic, which hit the van head-on. There were no seat belts, no guardrails.

All four of them died instantly.

My dad's memorial service was held on the old farm (what we now call the Founders Farm). About 2,000 people showed up. He had so many friends. Religious friends from many different backgrounds, and many speakers. The Allentown Band played.

My last interaction with him before he left for the trip was when I told him I wanted one of those big Russian hats. He smiled and said, "I need your hat size."

So I measured mine, wrote it on a Post-it note with a smiley face, "size 22," gave it to him, and asked, "Is there anything I can do for you while you're away?"

He looked up at me, pushed a big pile of papers toward me, and said, "Yeah, all of it."

In the suitcase that was sent back with him, there were three hats. One for each of his daughters.

After my dad passed, regenerative agriculture went mostly quiet. Meanwhile, organic farming was booming, especially with the USDA certification taking off. My father and the scientific results from the Farming Systems Trial were catalytic in our federal government passing the Organic Foods Production Act of 1990 (OFPA), which established national standards for organic agriculture in the United States.

The OFPA required the USDA to create a certification program, a labeling system, and a list of allowed and prohibited substances. This law aimed to ensure a consistent standard for organic products, protect consumers from misrepresentation, and facilitate interstate commerce in organic foods.

This act was passed by Congress on November 28, 1990. 68 days after my father's death in Russia.

My dad actually played a quiet but crucial role in helping bring about the USDA organic standards. Before they existed, certain groups had their own certification systems, and no one could agree on a unified approach.

He spent a lot of time meeting with these organizations and with government officials in Washington, trying to bring them together. He was a bridge builder, someone who could speak to both sides.

He didn't look the part. In jeans, hiking boots, and a bolo tie, he was once mistaken for a homeless man at the Waldorf Astoria in New York City. But he was humble, approachable, and deeply respected. And he enjoyed talking to everyone and anyone, no matter what their beliefs. The USDA even held a memorial service for him in D.C. after he died.

He believed in connection over division. He taught me the

value of staying grounded, talking to everyone, and never judging someone by how they show up, as long as they show up with heart.

When I got involved again after my dad's passing, GMOs were a major concern.

The Institute kept pushing regenerative, but it didn't have the spotlight or an accredited certification. I felt sad for my dad because "regeneration" as a term seemed to fade away.

But then one day (decades later), flipping through a Patagonia catalog, I saw it. The word *regeneration* boldly printed with a picture of a buffalo. It made me think, *maybe it's making a comeback.* And then, everybody started talking about it, thanks to Patagonia.

Patagonia catalog, Summer 2016

THE HISTORY OF REGENERATION:
How Rodale Institute Sparked a Movement

Since 1947, Rodale Institute has led the way in promoting farming that heals the soil, nourishes people, and protects the planet.

What began as a vision for healthy soil and food has become a global movement known as **regenerative organic agriculture.**

The Roots: J. I. Rodale (Maria's grandfather)

- **Founder of Rodale Institute**, inspired by Sir Albert Howard, a pioneer of natural farming.
- In 1940, J.I. and his wife Anna bought a worn-out farm in Emmaus, Pennsylvania, to experiment with organic methods like **composting, cover cropping, and crop rotation.**
- He observed that as the soil improved, so did his health.
- In 1947, he launched the **Soil and Health Foundation**—later renamed **Rodale Institute.**
- He famously declared in 1954, *"Organics is not a fad."*

The Innovation

Rodale helped promote key regenerative practices:

- **Cover crops**
- **Compost**
- **Crop rotation**
- **Eliminating the use of synthetic chemicals**

These methods rebuild soil fertility naturally and support healthier ecosystems.

The Next Generation: Robert Rodale

- After J.I.'s death in 1971, his son **Robert Rodale** (Maria's father) became the second-generation leader.
- A global thinker, Robert developed the term **"regeneration"** to describe farming that goes beyond sustainability to actively improve the environment and society.
- In **1989**, he co-authored the **Seven Tendencies Towards Regeneration**, foundational principles of the movement today.
- His vision redefined organic as **a holistic system** that renews **soil, ecosystems, and communities**.

The Science: Farming Systems Trial

- In **1981**, Rodale Institute launched the **Farming Systems Trial (FST)** to address a key challenge: the lack of practical, scientific evidence for farmers to adopt organic methods.
- For 45 years, the FST has compared organic and conventional grain systems using real-world practices and rigorous scientific analysis.
- It remains the **longest-running side-by-side trial of its kind in North America**, providing critical data on soil health, crop yields, resilience, and profitability.

One day, while my father was still alive, John Haberern brought a list of seven key findings from the first five years of the Farming Systems Trial. Right away, I noticed there was a correlation to several things we were working on at The *Regeneration Project*: the health, spirituality, and community areas. My father asked me to write up a list of how they were related. And that's how the *Seven Tendencies Towards Regeneration* was born. My father added the P words.

Since my dad's death, it's taken many years for the idea of regeneration to finally find a home in people's hearts and minds.

As regeneration spread, I was on my own journey of bringing regeneration to the Institute, in a whole new way.

At some point, I had to face an important truth: My job wasn't to win the approval of a parent who was gone or to carry out a mission that was never mine to begin with. My parents had their dreams but it became clear that my role was to live out *my* purpose, and to make sure I didn't pass down the emotional weight I'd inherited to my own children, which is why I worked hard to make the Institute as independent as possible, and also give my children the freedom to make their own life choices.

To me, regeneration isn't just about soil. It's about healing patterns, restoring trust, and building something that can outlive any one person.

That's what I set out to do with the Institute. Behind the scenes, I worked to make it independent, endowed, and no longer reliant on family involvement. That's a legacy I'm proud of.

The older I get, the more I see the same cycles in society repeating. Every generation grows up thinking they've discovered something brand new—whether it's fermentation or regenerative agriculture—and we say, "Welcome. Let's build a new future together." The challenge is making sure each new wave doesn't erase the one before, but strengthens it. And that older generations leave a map and a guidebook, when possible.

Seven Tendencies Towards Regeneration
(Reimagined)

1. **Pluralism**
   - Greater diversity of plant species and ecosystems
   - Broader representation of people, cultures, and ideas in communities and businesses
   - Richer personal experiences and openness to new ways of thinking and being
2. **Protection**
   - More plant cover reduces erosion and supports thriving soil microbiomes
   - Economic and cultural resilience through inclusive, diversified systems
   - Strengthened physical and emotional well-being, fostering personal resilience
3. **Positivity**
   - Chemical-free growing methods support flourishing soil life
   - Clean air and water contribute to better public health outcomes
   - Letting go of harmful habits or thought patterns creates space for joy and growth
4. **Permanence**
   - Deep-rooted perennials stabilize and enrich the landscape
   - Stable, values-based businesses strengthen communities long-term
   - Grounded spiritual practices foster lasting personal meaning and connection
5. **Peace**
   - Regenerative systems reduce the need for disruptive inputs like pesticides
   - Community-centered approaches can lower crime and increase safety
   - Emotional healing fosters compassion, empathy, and understanding

> 6. **Potential**
>   - Nutrients accumulate near the surface, ready to nourish life
>   - Resources circulate more equitably, enabling broader access to opportunity
>   - Human and ecological potential flourishes when barriers are removed
> 7. **Progress**
>   - Improved soil structure boosts water retention and long-term fertility
>   - Community life thrives through inclusive health, wealth, and participation

That's where Rodale Institute is now. Regeneration has gone beyond the soil and the farm; it encompasses all of nature. It's woven into how we lead, how we listen, and how we plant seeds for the future."

Maria set the stage. Because of her vision, I was able to step into something much bigger than myself. When I stepped into the leadership role at Rodale Institute, it was more than a career move. For me, it was a calling.

I officially accepted the job in May of 2017, with a start date set for July 1. I wanted to give my former employer, Rodale Inc., a full two months' notice. It felt like the right thing to do.

Jeff Moyer, the CEO of the Institute at the time and the person whom I would eventually replace upon his retirement, supported my transition plan. But he added, "There's a meeting happening at the Institute in about a month, and you *need* to be there. Take a vacation day, do whatever you need to do—but you don't want to miss that meeting."

So, I agreed.

The meeting was set, and the night before, a group of people—many of whom I didn't know—flew in from all over the United States. We had dinner at a restaurant called Glasbern. I met people like Will Harris (regenerative rancher and founder of White Oak Pastures), David Bronner (CEO of Dr. Bronner's), and Rose Marcario (former Patagonia CEO), along with leaders from social fairness and animal welfare movements, as well as leaders from Rodale Institute.

Jeff Moyer gathered us around the restaurant, which we'd taken over for the evening to read aloud the *Seven Tendencies Towards Regeneration*. That night, over dinner, I learned about a major fallout that had happened back in March at Expo West in Anaheim, California, the biggest trade show in our industry. There had been a massive argument over the word *regenerative*. It sparked a long, contentious email thread that quickly became ugly.

But instead of letting it spiral, Dr. Bronner's, Patagonia, Rodale Institute, and others harnessed that energy and turned it into something positive. They decided: we need to put a stake in the ground around the term *regenerative*. Let's launch a new certification. One that builds on the long-standing efficacy of organic.

Rodale Institute agreed to host the initial summit: a two-day workshop at the Institute. We gathered around a big conference table and started creating what is now known as the Regenerative Organic Certification.

Sixty-eight of the world's top one hundred food companies claim to have a "regenerative strategy" for their supply chain and the farmers within their supply shed.[3] But the word regenerative means *nothing* on its own.

Without a standard, without accountability, it's just marketing hype.

That's why the certification, Regenerative Organic Certified®, matters. It puts real meaning behind the word and real impact behind the movement.

The certification is grounded in clear principles that define what "regenerative" actually means. These tenets echo the original vision of Robert Rodale.

To be Regenerative Organic Certified®, a farm must meet four fundamental standards: it must be certified organic, improve soil health, support social fairness, and uphold high animal welfare practices.

Regenerative organic products must first be certified organic under an accredited certifying body. That means eliminating synthetic inputs like herbicides, pesticides, and chemical fertilizers. You might see terms like "regenerative conventional" or "regenerative non-organic" but they are contradictory and not truly regenerative. Any system that relies on synthetic inputs or chemicals is inherently extractive and cannot truly regenerate. It must be organic, first and foremost. From there, the certification emphasizes soil health because healthy soil equals healthy people.

Regenerative organic farming restores and rebuilds soil, rather than depleting it. The certification also includes strong standards for social fairness, like making sure that farmers and farmworkers are paid fairly and treated with dignity. It also requires the humane treatment of animals.

Together, these four pillars offer a clear, science-based, and values-driven definition of regeneration—one that moves beyond buzzwords. Plus, the Regenerative Organic Certification is third-party verified through rigorous, annual audits by 15 trusted certifiers, ensuring transparency, integrity, and trust.

What's critical to understand is that we're not abandoning the pioneering work of Robert Rodale. We're building on it. Regenerative Organic is the next evolution, setting a higher bar for soil health, animal welfare, and social fairness. In fact, the Regenerative Organic Alliance (ROA) was founded, in part, as a response to the USDA's decision to allow hydroponics under organic certification, a move many felt compromised the integrity of organic standards.

As the CEO of the Regenerative Organic Alliance, Christopher Gergen noted, "Building and maintaining trust in the marketplace is critical. The term "regenerative" is often used loosely, without a clear definition or third-party verification, which can erode consumer trust. That's why a standard like Regenerative Organic is so important: it offers a credible, verifiable benchmark, builds confidence, and supports the education and awareness needed to drive demand for truly regenerative farming."

If we want regenerative to mean something, we have to be willing to talk about the harder stuff too, like tillage. There's a concerning narrative out there that tillage is the primary culprit of soil degradation and climate change. And while excessive or misused tillage can disturb the soil, the real culprit isn't the act of tilling. "It's the chemicals in our farming systems that are leading to all of the negative outcomes that we know," Christopher Gergen explained.

Tillage is actually a tool. In Rodale Institute's Farming Systems Trial, we've discovered that it's not tillage alone that depletes soil health. It's tillage combined with chemical use.[4] In fact, the healthiest soils in the trial belong to organic plots, where tillage is used carefully and strategically, not as a default. We use tillage surgically, meaning once in a while, we need to go in and till to establish a new crop. This can actually make the soil healthier over the long term.

Some groups, often funded by large chemical companies, have pushed an anti-tillage agenda under the guise of regeneration. In this version of the story, no-till becomes synonymous with sustainability. But instead of reducing harm, it often means replacing tillage with chemical sprays to terminate cover crops. That's not regeneration. That's just a new version of extractive agriculture.

The good news is we've seen remarkable growth in the Regenerative Organic Certification (ROC). Nearly 2,700 ROC certified products are on the market from 330 consumer packaged goods companies, including major brands like Lundberg, SIMPLi, Dr. Bronner's, and Whole Foods.[5]

Nearly 20 million acres and over 70,000 ROC certified farms exist across 46 countries. That's remarkable growth considering the first meeting on regenerative organic happened in 2017.

Within the first year, the Regenerative Organic Alliance was formed, and the standard officially launched in 2020. In just five years, we've seen explosive growth and expansion in certified farms, acres, products, and brands.

Interestingly, another turning point happened around the same time: the global pandemic.

In 2020, we witnessed something profound. People started asking more questions about where their food came from and how it was grown. Faced with uncertainty, people instinctively reconnected with the land.

Direct farm sales (people like you and me purchasing food from farmers in our community) jumped 35 percent as people began sourcing food directly from farmers in their own communities.[6]

Nearly 18 million new gardens were planted across the United States that spring.[7] Seed sales soared. Canning supplies flew off the shelves. CSA participation surged. Urban gardening doubled.[8]

It was as if something ancient was activated: a biological reflex to get outside, touch the soil, and feed ourselves and our families better.

This wasn't just about food. It was about grounding. We know from neuroscience that touching the earth can literally change our brain chemistry. And suddenly, millions of hands were in the dirt.

At a deeper level, something was happening: we were remembering.

Remembering that health begins with the soil.
That food is medicine.
That community matters.

Even as we face human and environmental crises of massive scale, this is where my hope lives: in people waking up. In the quiet power of regenerative organic farmers who steward the land with care. And in the growing movement of people choosing to support those farmers. Not just for food, but for health, healing, and the protection of our shared future.
As more and more people join in, so are our social systems. School districts and healthcare systems are asking, "What's the best food we can serve our communities?"

I can't help but think of Robert Rodale. The man who believed that organic was just the beginning. He envisioned a regenerative food system, one based on reciprocity, not extraction. On resilience, not exploitation.

If he could see the global momentum, the shift in our collective awareness, the farmers rebuilding their soil and their communities, I think he'd be proud.

Not because we've arrived, but because we're gaining momentum as we keep moving in the right direction.
That's how regeneration works.

A regenerative organic future isn't just a dream. It's already taking root.

As more people recognize that regenerative organic food is

not only free from harmful chemicals but also richer in nutrients that nourish our bodies and protect our families, the demand grows stronger.

We're planting the seeds today for a healthier, more resilient tomorrow.

One person, one meal, and one farm at a time.

Andrew Smith, PH.D, Chief Scientific Officer, Rodale Institute

Chapter 9

THE LONG GAME

Organics is not a fad.
It has been a long-established practice -
much more firmly grounded
than the current chemical flair.
Present agricultural practices are leading us downhill.

—J.I. Rodale, 1954

You have to study nature for a *long* time to understand how it works.

That's where science comes in.

For 45 years, Rodale Institute has been running the Farming Systems Trial, the longest-running side-by-side comparison of organic and conventional grain farming in North America. The purpose of this long-term study, originally called the Transition Study, was designed to document soil health and crop production through the transition from conventional to organic. The goal for this science is to develop solutions that uncover the barriers to transitioning to organic and develop solutions that make the shift easier and more profitable for farmers by harnessing the power of nature.

It's not just research for research's sake. It's about giving farmers the tools, data, and knowledge to make the switch successfully.

Dr. Andrew Smith has been leading this groundbreaking research at Rodale Institute for about a decade. He is our chief scientific officer.

Drew's career in the science of organic farming was motivated by a troubling trend that he stumbled upon in college. Around that time, he picked up a book by Rodale Institute collaborator Dr. David Pimentel, who laid out the stark reality: despite decades of chemical use, we'd made little progress in reducing crop losses from insects and disease, and yet we'd flooded the environment with synthetic inputs. In 1997, the U.S. hit its highest recorded pesticide use since tracking began in 1960, and the numbers kept climbing.[1]

At the time, farmers were applying just under 900 million pounds annually of active ingredient insecticides. Today, that number has soared. Herbicide use alone now tops 800 million pounds, with total pesticide use likely well over 1.2 billion pounds per year.[2]

"How is that a sustainable and healthy approach for the future of humanity?" Drew asked himself.

Drew had a degree in Agronomy, but all the jobs available in the agriculture industry involved selling more chemicals to farmers. That didn't resonate with him. He wanted to do something different. Something that aligned with the kind of future that he believed in.

So, Drew joined the Peace Corps and served in Guatemala for two years, where he saw firsthand how widespread chemical use in agriculture had become. In some communities, Drew witnessed female farmers working with children strapped to their backs while nearby husbands or sons sprayed chemicals on the crops. Often, those same crops were later rejected for export because the chemicals used to grow those crops weren't approved by the USDA and FDA, leaving farmers unpaid and vulnerable to exploitation by middlemen, or *coyotes*, who offered cash but at unfair prices.

That experience reinforced Drew's conviction: agriculture needed to break free from the chemical treadmill. He went on to earn a master's degree in integrated pest management, and eventually, he and his wife started their own certified organic farm, which remains their home farm today.

But their farm struggled financially, and they couldn't make the business of the farm pencil out in the early years. So, Drew pursued a PhD with the hope of gaining even more skills and income to run a successful organic farm. Not unlike over 60 percent of farms that rely on off-farm income.

Through his research studies, Drew ended up being hired by Rodale Institute to lead their long-term science, driven by his growing concern over the unchecked use of chemicals in agriculture as well as his desire to help other farmers to farm organically in a more profitable way.

As he said, "At Rodale Institute, we're trying to find solutions so farmers can transition to organic."

That transition, however, takes time and so does the science behind it.

Short-term studies simply don't capture the full picture. As Drew explained, "From a scientific perspective, we have a real struggle. Most university studies last two to three years, maybe five at most, given the length of the average PhD program. But almost every soil scientist will tell you that any major change to the soil, burning a forest, clear-cutting, tilling, or switching from conventional to organic, takes four to five years before you even begin to see measurable change in the biology.

Think about it. You and I don't change that much over the course of a few years. But if we ran into a childhood friend that we have not seen since graduating from high school during our adult life, we might not recognize them.

In a short-term study, you're simply not going to see change over time and that's a problem. Too often, studies compare organic and conventional methods over just a few years and find that organic yields are lower. But they're not giving the soil time to change biologically, which it absolutely does, especially after five years, and even more so over decades. The first meaningful changes in FST occurred in year 5. Yields were equal between organic and conventional and there were measurable differences in soil health, with the organic portion having higher organic matter, also known as soil carbon.

It's not just about needing long-term research; it's also about taking a systems approach, which is what Rodale Institute is all about. Most scientific studies are reductionist. They isolate a single variable to say, 'This caused that.' But that's not how farming works in the real world. Farmers aren't isolating variables; they're managing whole, dynamic systems. And that's exactly what the Farming Systems Trial is designed to reflect.

The Farming Systems Trial allows us to evolve over time. What we're doing today in the field isn't the same as what we did in 1981, when the study began."

We don't know what questions the future will bring. If a study stops, it takes another five years just to start seeing meaningful data again. While conventional agriculture has become reductionist, organic systems are interconnected. That's why we need long-term systems trials, to understand how this complexity plays out in soil health, water quality, and more.

That is the value of long-term research. It adapts and morphs. It plays the long game, tracking slow, meaningful change.

The most important insights only emerge after years of careful observation. And when they do, the findings can be game-changing.

In the first year of the Farming Systems Trial, researchers discovered something surprising: the organic soybeans actually outperformed the conventional soybeans. But the organic corn struggled. That's because corn needs a lot of nitrogen to grow, and in an organic system, it takes time for the soil to build up that nutrient naturally. Soybeans, on the other hand, can make their own nitrogen, so they didn't have the same problem. But with the right crop rotation and a little patience, by year five, the organic corn was yielding just as much as the conventional. That showed it takes about four to five years to naturally rebuild the soil's fertility.

That insight improved how Rodale Institute advises farmers today.

Another major finding came from studying ten years of rainfall and crop data. In eight of those years, there was little rain, and one year had a severe drought, which is probably why they decided to look back and analyze the data. In the early drought years, organic and conventional systems performed about the same. But after a few years, the organic crops consistently did better in dry conditions.

Why?

Because organic soils hold more carbon, which helps them retain water and stay resilient in both drought and heavy rain.

In fact, during heavy storms, water infiltrates into organic fields faster (more than two inches per hour) while conventional fields absorb less and are more prone to erosion. A month later, when the rain was gone, the organic soil still had water stored deep down, giving plants more time to thrive.

Why the Rodale Institute Farming Systems Trial Exists and Why It Matters

The roots of the Farming Systems Trial go all the way back to **1939** in Cheshire, England. At a pivotal meeting, agricultural scientist **Sir Albert Howard** warned that artificial fertilizers were depleting soil health. He shared early observations that compost-fed plants were stronger, more resilient, and healthier overall.

Then physician **Robert McCarrison** presented striking evidence that connected diet to health outcomes. After studying different communities in India, he showed that populations eating nutrient-dense food enjoyed far better health than those relying on nutrient-poor diets.

The message was clear: **how food is grown directly affects human well-being.**

Everyone agreed a direct comparison between natural and chemical farming was needed. But only one person followed through: **Lady Eve Balfour.** She launched a long-term study and documented the first few years in her book *The Living Soil*. Her study continued well after the book was published, running for around 30 years before concluding, but it lacked today's scientific standards, like replication and controls, and as chemical agriculture took off, her work was largely dismissed.

Fast forward to 1981. Seeing the gap in credible, long-term research, **Robert Rodale** founded the *Farming Systems Trial*. He insisted it meet the highest standards of scientific rigor so its results could be published, respected, and replicated. Robert also helped popularize the term **"regenerative agriculture,"** now central to conversations about soil health.

Today, the Farming Systems Trial is the longest-running side-by-side comparison of organic and conventional farming in the U.S. and it's still going strong.

And here's where it gets a bit scientific but important: organic matter in soil holds water more loosely than the soil particles themselves, meaning plants can still access water even when things start to dry out. This helps delay the point when plants completely wilt from lack of moisture, giving them a better chance to survive during dry spells.

In a world where weather is becoming more erratic and less predictable, with floods one year, droughts the next, these findings show that organic systems are better equipped to handle the ups and downs. They store more water, soak up rain more efficiently, and help crops grow through tough conditions.

That's a big deal for farmers and for our food system.

One of the most exciting outcomes of Rodale's research is how it has inspired others. After Rodale Institute launched the Farming Systems Trial in 1981, a wave of other trials started across the country. Many people questioned whether what worked at Rodale Institute in Pennsylvania would translate to places like Wisconsin, Iowa, or California.

The answer was a resounding YES. Overwhelmingly, across the board, soils improved when farmed organically, regardless of location.

At Rodale Institute, we decided to explore nutrient density, meaning the actual levels of vitamins, minerals, and beneficial compounds found in the crops. To find out how to keep these nutrients in the soil and our food, we launched the Vegetable Systems Trial in 2016 (and we also began nutrient testing in the long-running Farming Systems Trial).

What we found so far is interesting and a little surprising.

Even though the mineral levels in organic and conventional crops aren't dramatically different, the exciting differences are showing up, just in other ways.

Organic crops have been found to contain more protein, more vitamins, and higher levels of beneficial compounds like lutein, lycopene, and beta carotene. These nutrients are now getting serious attention from doctors and health experts for the role they play in keeping people healthy and helping to safeguard them from chronic illnesses such as cancer.

It's not just about how much magnesium is in your vegetables. It's about how the *entire* nutritional profile of a crop changes based on how it's grown.

And that brings us back to the original question that launched Drew into this work:

What are the long-term effects of the chemicals we're using to grow our food?

The evidence is piling up and it's alarming.

Chemicals aren't just harming the environment; researchers have examined their presence in human bodies, too.

Biomonitoring studies conducted by the Centers for Disease Control and Prevention (CDC) have reported measurable atrazine metabolites in human urine samples, indicating ongoing exposure at low levels.[4]

Not only are we still using chemicals known to be harmful, we're using them on an even more massive scale.

But it doesn't have to be this way.

The Farming Systems Trial and other research at Rodale Institute has proven that a better way is possible.

So why aren't we moving in that direction faster?

Because this is a long game. The science, the innovation, and the urgency are all moving in the right direction, but to truly innovate takes time.

Our goal is to get more farmers and people like you and me to ask:

Why are we even spraying chemicals at all?

Maybe that question leads more farmers to take the leap to organic. When they're ready, Rodale Institute is here to help them cross that bridge.

The die-hard organic farmers, like some we'll meet in the next chapter, have already made the switch. Now the goal is to bring more farmers and everyday people along too, which brings us to the ultimate question:

Is going organic really worth it?

One of the key findings of the Farming Systems Trial says yes. Organic systems proved to be more profitable than conventional ones. Yes, organic had higher labor costs, mainly due to more hands-on fieldwork, but those costs were more than offset by price premiums. This trend has held true across numerous studies: even when organic is more expensive to operate, it consistently comes out ahead financially.[6]

The insane news? Even without price premiums, the regenerative organic system was *still* more profitable. In other words, if the whole market went organic tomorrow and premiums vanished, organic would still come out ahead.

That's a game-changer. And the economic benefits go beyond profit margins. In organic systems, more money stays in the local economy.[7] That's because labor is often done by the farmer, family members, or neighbors, not outsourced to global input companies.

Dr. Edward Jaenicke, Ag Economist at Penn State University, and his graduate student Julia Marasteanu did the work on an important study. The Organic Trade Association found their study and turned it into a public report. Why were they so intrigued by what one study had to say? Because it found that counties with higher per capita numbers of certified organic farms had stronger local economies.

So what does all of this mean for you and me?

The choice for us is simple: buy organic when you can. Not just for your health, but to support the farmers in our communities who are doing the courageous thing.

Every organic purchase is a vote for a better, more regenerative food system.

And it's already happening. The shift is no longer theoretical. It's visible on grocery store shelves, in supply chains, and in the choices people are making every day.

This got me thinking. Why are more people buying organic food? What is *really* causing this inflection point for organic? And what can we learn from the data about where things are headed?

To explore that, I spoke with Jay Margolis, CEO of SPINS, a company that's been tracking data on the natural and organic marketplace for more than 25 years. SPINS was built to help brands and retailers understand what's happening on store shelves, from what consumers are choosing to the values driving those decisions.

They work with thousands of brands including household names like Dr. Bronner's, Lundberg Farms, and Amy's Kitchen, as well as emerging innovators, like Purely Elizabeth and Califia Farms. The brands they support are usually the ones leading innovation and setting the trends that others will follow. Even the giants, like General Mills, Pepsi, and Mondelez, turn to SPINS to understand where the market is going next.

If you want to know what's shaping the future of food, you have to follow the trends and the data behind them.

"The rise of certifications in the food system is one of those trends. From a consumer perspective, certifications are essentially symbols of trust," explained Jay. "When you're standing in front of a shelf packed with 18 different cracker brands, you're looking for something that meets your needs, whether that's dietary preferences, ethical values, or environmental concerns. Shoppers have consistently shown they're willing to pay more for that peace of mind, for themselves, for the planet, and especially for their children."

Jay backed this up with some compelling data. "In the past

year alone, sales of Regenerative Organic Certified® (ROC) rose about 22 percent. And while sales of other sustainable certifications (e.g., organic, fair trade) are also outpacing the market, the addition of a ROC certification further accelerates that product growth."

Even though ROC is still in its early days, it's quickly emerging as a major force in the industry. It reflects where things are headed, not just in farming, but in how people are choosing what to buy and what brands to trust.

Jay went on to explain, "As the role of primary grocery shopper has shifted from Boomers to Millennials, a new pattern is emerging. Millennials tend to be more aware of what they're buying and what it represents. And as Gen Z steps into their own as consumers, that trend is deepening. Though they're still early in their grocery-shopping lives, Gen Z is already selective about the brands they support. More than just choosing products, they're choosing identities. They want brands that reflect who they are and what they care about. There's been a clear generational shift.

Gen X and Boomers were drawn to fashion as a form of self-expression, with things like designer bags, a particular pair of shoes, or a well-known label. Those were identity markers. But Gen Z has flipped that script. They're more likely to wear a monochromatic sweatsuit with no visible branding and post a photo of themselves holding a can that says *Regenerative Organic Certified*."

Food has become a centerpiece of identity, conversation, and social currency. It's featured in TikToks, Instagram stories, Substack posts, and countless online communities. Young people aren't just eating food - they're talking about it, analyzing it, and turning it into a social movement.

Gen Z is a generation that knows they're inheriting a less healthy planet and many feel a responsibility to do something about it. They see food as a way to take action. One of the most powerful tools they have is how they spend their money.

As Jay put it, "I vote with my wallet every day at the grocery store. And I believe this generation is doing the same. As an individual, it's hard to change national policy. But you can influence what happens on a farm two years from now just by choosing organic or regenerative organic products."

The food system, in many ways, has become ground zero for social change. And younger consumers seem to understand that, intuitively.

There's also something else that happens. Something almost universal. People may spend years buying conventional products without giving it much thought.

But when they become parents, everything changes. Suddenly, the same person who always grabbed the cheapest gallon of milk from the shelf is reaching for organic.

And that shift doesn't stop with milk. It carries over across dairy, into meat, produce, and snacks. "Suddenly, entire grocery lists begin to transform. Shoppers spend years being aware of the benefits of organic or natural products, but it is not until they are making decisions for their family that we start to see patterns shift. They're willing to spend more on the products they believe are better. What we've seen is that these purchases become gateways. A parent might first buy organic for their child, but eventually, it changes what the whole family eats," Jay noted.

A single choice at the grocery store becomes the start of a broader transformation.

And it's not just what's *in* the product that matters. Packaging is another frontier of change.

Jay pointed out that both consumers and retailers are increasingly seeking out containers made from recycled or upcycled materials, using soy inks instead of petroleum-based ones, and moving toward compostable options.

"One of the more interesting things we found," he said, "is that across multiple beverage categories plastic use is rapidly declining, and in response, materials like glass and aluminum are going up. Across two generations plastic has gone from a marvel of the industrial age to a symbol of planetary disregard."

If shifting consumer preferences are changing the future of food, those preferences have implications on the way our land is farmed. Sometimes in the most surprising ways.

THE FARM IS HERE

Farmer Dave Marshall

Chapter 10

SO MANY DAVE'S

JEFF TKACH

THE FARM IS HERE

Last fall, I made a farm visit that I'll never forget. On a beautiful October morning, I rolled up to Dave's farm in central Pennsylvania.

I'd heard about this guy.

How this conventional farmer made the switch to organic and was now growing organic grain at scale. This guy had a reputation for being hyper passionate about organic farming.

As I drove up to his farm, the first thing that I noticed was the political sign out front. The visit was at the height of political tension in the U.S., just before the 2024 presidential election. When Dave first greeted me, pulling up on his Kubota RTV, he was wearing a well-worn black T-shirt and jeans, a Carhartt baseball cap, with a sturdy build and a big, gregarious smile.

Dude drove big tractors.

He embodied the classic image of a farmer, fitting all of the familiar stereotypes.

He was waiting in anticipation for my arrival and welcomed me to his farm, though there was a hint of speculative reluctance. I was on my way to a family function later that day and was dressed for the occasion.

Perhaps my appearance was cause for his suspicion?

But within moments, we hopped on the Kubota and went for a ride around the farm. Once I got him talking about organic farming, he disarmed.

Dave was also getting his political digs in as we drove along the farm and made it very clear where he stood on that matter. He did not hold back on that topic one bit even though we had just met.

I found that interesting.

But the beautiful part was, he made no bones about the fact that organic farming is for everyone.

As outspoken as he was about his political points of view, he was even more outspoken about his position on organic farming.

Dave started telling me about his own journey. His background was in forestry. He started a logging business and cut down trees for a living. After sustaining one too many injuries from trees falling on him, he pivoted to farming. He started buying farmland in rural Pennsylvania, initially farming conventionally, but quickly saw it wasn't very lucrative.

Then he started learning about organic, dabbling in it, and realized how much more money he could make. He got really into it, researching techniques for weed management and crop rotations. He tried these techniques on his own farm and was astonished by the power of nature to control weeds naturally and by how healthy his soil became once he introduced these new practices.

He drove me around his farm on this beautiful fall day, showing me each field, how he was managing weeds through cover cropping, and pointing out areas where he was improving soil health.

At one point, we were standing at the top of a hill looking down on one of his fields and a pheasant flew by. Seeing a pheasant in the wild is very uncommon in South Central Pennsylvania these days because chemical farming has nearly wiped out the pheasant population.

But there were pheasants on Dave's farm. He was so proud of that, along with all of the other biodiversity that was flourishing since he began farming organically.

As we overlooked the border between his farm and his neighbor's, we could see his neighbor had planted conventional corn which was lagging in growth compared to Dave's corn in an adjoining field.

He pointed out the difference between the two fields, saying, "Look over here, now look over there." Dave's organic corn was clearly healthier. Anyone could see that with the naked eye. It was obvious. There was no data behind it, just this visual anecdote that made an impression. He even said, "My neighbor's jealous of what I've got going on in my fields." There was a lot of pride in that moment, and rightfully so.

We spent the whole morning together. He introduced me to his daughter who will take over the farm someday, along with her husband who came to organic farming through Dave. They're raising beef cattle and exploring the idea of raising organic poultry in an effort to build diverse income streams to supplement their grain business.

There's a plan to transition the business to them, and Dave is very focused on what legacy he is leaving to his family. That was a big reason why he transitioned to organic. He doesn't want to leave a financial burden; he wants to leave a financial opportunity.

Upon first impression, Dave is a guy you'd least expect to

enter organic farming as passionately as he has.

But the deeper you look inside the other, there you find yourself.

That's what I experienced on that beautiful October day with Dave.

We connected through our shared love of organic farming and all that it represents. We related through our mutual commitment to land stewardship, healthy food and healthier ecosystems.

Dave is passionate about growing good food, and so am I. He cares deeply about stewarding his land, about leaving the farm to his family in a healthy, profitable way. He's passionate about building a resilient food system for his community, about soil health, biodiversity, and farm diversity.

I could go on and on, but the point is: we connected.

The political sign out front. I no longer saw it as a polarizing symbol.

Dave and I found common ground.

When I think about who is driving the demand for regenerative organic food, that person likely has very different political views from the person that is *actually* producing the food.

When the 26-year-old tech entrepreneur from San Francisco walks into a Whole Foods to purchase a certified organic chicken, she is relying on farmers like Dave in rural America, growing certified organic grain to feed that chicken.

Everything is connected to everything else.

I have visited so many farms over the last eight years and I am endlessly amazed by how many Dave's there are out there. Meeting farmers like Dave always seems to hit me like a giant paradox.

And it always seems to go like this. We found common ground.

We are united in the soil.

But not all farmers come from rural America.

Christa Barfield, CEO of Farmer Jawn Agriculture,

After 10 years of working in the healthcare industry in Philadelphia, Christa Barfield decided to leave her job and go to Martinique, an island in the Caribbean where she has traced her ancestral lineage.

When she decided to go on this vacation, her intention was to rest and recover from her draining life in corporate healthcare. It was her first solo trip, her first time out of the country — and by chance, the rental property that she stayed in was owned by Black farmers. That's how she knew it was destined.

While she was there, she spent time on the farm, got to know the farmers, and touched soil for the first time in her life.

By the end of that trip, she thought *I want to go home and start a farm.*

In many ways, Christa didn't find agriculture; agriculture found her. Her story deeply intrigued me. So I spoke to Christa about her journey into agriculture and that trip that changed her future.

She said, "Meeting Black farmers who were stewarding their own land and selling food, sometimes to people who didn't even look like them, opened my eyes. I realized this is a responsibility we all share: to make sure everyone has access to good food."

When she returned to Philadelphia, she started trying to figure out what this new calling meant for her. In order to sustain herself financially, she created a routine: from 5 a.m. to 9 a.m. and again from 5 p.m. to 9 p.m., she worked gigs, like Instacart, to make money.

In the middle of the day, Christa focused on building her

farm. She started volunteering at farms whenever she could. She realized pretty quickly that tasks like packing vegetables or harvesting weren't going to teach her the bigger picture of how to run a farm.

That's when she leaned into her background. For years, she'd managed medical practices, and her responsibility was, essentially, to help doctors manage their finances. She took her skills, combined them with her passion, and stepped fully into becoming a farmer.

Her very first farm started in her backyard in Germantown. It was just 24 square feet—a friend bought her a small, six-by-four greenhouse and set her up with the basics: a rig, irrigation using the garden hose on a timer, and heating equipment.

Christa had no formal training, no tutorials, and didn't do any deep dives into agricultural books, or YouTube. She just followed her intuition, which she now understands was really ancestral knowledge coming through her. She began by growing herbs. Her background in healthcare inspired her to apply what she knew about healing and medicine to the natural world. So, she started a tea company. She infused the herbs she grew into locally sourced honey and began selling the products. To her surprise, the response was immediate.

At Christa's very first public event, someone approached and said, 'You should sell your products in our stores.' She was doubtful. She thought, *I've only been farming for five minutes. How could I be ready for that?* But this stranger insisted. And before she knew it, her teas and infused honey were on shelves at specialty grocery stores across Philadelphia.

From there, she had momentum.

Christa wasn't raised around farms. Growing up, she was a Girl Scout and loved camping. But beyond the outdoors, she never really interacted with plants.

She grew up in the northwest section of Philadelphia, with one main street: Germantown Avenue. This cross-town road runs through many neighborhoods in Philadelphia, including Mount Airy, Chestnut Hill, and Germantown. In her words, "As you move through these communities, you can literally see the changes in food access, food insecurity, and nutrition deficiency. The makeup of each community shifts, and the biggest factor is food."

In 2020, with her farming business on the rise, Christa started going to corner stores in her neighborhood. She asked the store owners, 'Would you be willing to carry food from a farm? Why don't you sell produce?'

That's when it hit her. To heal a nation, we have to feed it well. "When people say *food security is national security*, for me, it feels much more personal, right at home," Christa explained. That's been her mission: to use her farm to feed people well.

In March 2020, she leased a one-acre lot with 40,000 square feet of greenhouse space, just 15 minutes from where she grew up.

From that first backyard greenhouse to a small community garden plot, to suddenly running a large greenhouse operation, the scale of her work grew rapidly.

That was when her farming company, Farmer Jawn, really began to take off. She started building a CSA community, people across the region got to know the organization, and soon enough, she was on the news and in print everywhere.

As Christa explains, "This was the moment I truly became a farmer; when my passion and purpose came together with opportunity in 2020."

And that's when Rodale Institute entered her life.

Sam Malriat and a few team members came out to her newly expanded farm - the one-acre site with 40,000 square feet of greenhouses. Sam was always a phone call away. If Christa was having issues, the Rodale Institute consulting team showed up. They helped her fill out paperwork, provided technical assistance, access to equipment, and were ever-present as her farm expanded.

At the time, regenerative organic was just starting to gain attention. But for Christa, she'd never known how to farm any other way. "I didn't understand why anyone would use chemicals—why would I buy them in the first place?" Christa explained.

"Rodale has always felt like family, and that's exactly how they treated me. My farm is better because of the relationship I've built with them, and I'm better for it too. I'm stronger, more knowledgeable, more tapped into not only how the industry is shaped but also where it's going — because Rodale is the future. And I'm proud to be part of that," Christa said.

For Christa, farming is really about healing.

Over time, she developed what she calls the *Farmer Jawn Formula*, which at its core, is about health. Christa believes we can measure health in many ways. Most people focus on environmental health or physical health; the two pillars she always stands by: people and the planet. But in her eyes, the third pillar, the one that connects everything, is *social health*.

"Social health often gets overlooked. Yet food is the ultimate connector. Food has the power to bring us together across all those pillars,' said Christa. 'And what holds everything up, people, planet, and social health, is the soil. That's the foundation. Right now, our soil is suffering, contaminated by forever chemicals and harmed by shortsighted policies. From where I stand, leaders aren't prioritizing it in the way they should. But I still believe change is possible. I believe that the work we're doing now, at this moment, is helping to shift the future. I want my children, and my children's children, to live in a different world when it comes to food. Because food is medicine. It has real healing properties. And that's why I say: to heal a nation, we must feed it well."

Her call to heal through food suddenly felt even more personal when she learned that farming had deep roots in her own ancestry.

Throughout her life, Christa had spent time in North Carolina. As a child, when she visited her mother's family, the land they had once owned was already gone. After a disaster in Tarboro and Rocky Mount, where the Tar River flooded, companies came in, bought up land, and sold it off. By the time she started visiting her grandfather around age eight and into her teens, he was already living in an apartment. Christa only knew him as a mechanic, never as a farmer. She had no idea about her agricultural roots.

Even when she attended an all-Black private school, where slavery and its deep connection to agriculture was openly discussed, she still didn't make that personal connection.

It wasn't until two years ago when she was talking with her 97-year-old grandmother, her father's mother, that she learned the truth. Her grandmother was telling her stories about her siblings and other relatives in North Carolina. And she casually mentioned, "You should come to the family farm. We have land."

So last year, Christa got to walk for the first time on the Barfield family land. She stood at the Barfield gravesite where generations of her family were buried. She was there with her own children. It was emotional, overwhelming, and beautiful all at once.

By that point, she had already been farming for several years. But to find out that her family still owns 200 acres of land in North Carolina. It was a revelation.

That experience confirmed what she had already felt deep down: agriculture didn't just find her by accident. It called her.

"I've come to see myself as being in service to the soil. And being in service to soil means I'm also serving people, seeking to change lives for the better. That belief grounds me in my purpose," said Christa.

The weekend before we spoke, Christa attended an event hosted by *WURD Radio* that reminded her of that purpose in a powerful way.

WURD Radio runs an initiative where they pick a Black-owned business in the community, announce it to their listeners, and rally the community to show up in support. They even provide a bus for transportation.

On Saturday, they chose Farmer Jawn, located 45 minutes outside of Philadelphia, which can make access a challenge for many urban residents. But that day, over 400 people came in just three hours. "The line stretched long and it was overwhelming in the best way. For me, the bigger message was about access. There are people who don't look like me who can come shop anytime - they have cars, they have the means. But there are so many others, people who had heard of Farmer Jawn but had no way to get here — until that bus pulled up. Being able to share with them the importance of eating well, of knowing your farmer, of understanding where your food comes from, that was transformative," Christa explained.

"What struck me most was how the message resonated with everyone in that room, regardless of background. That's when I know this work is real, that it matters. When you see people light up with the realization that food connects us all, that's when you know it's impactful. That's when you know it's a catalyst for change."

As I see the movement Christa has created in Philadelphia, I believe it will take root in every city around the country because the truth she carries is universal:

Food heals, and soil connects us all.

Steve Yelland is a member of the Warehime family. The Warehime's are known for their legacy in the snack food business, with close ties to Snyder's of Hanover and Hanover Foods. Over the decades, they've not only shaped regional food production but also invested heavily in preserving farmland and supporting the local community in Southcentral Pennsylvania.

Steve's family aren't newcomers to agriculture. Their roots run deep, back to the early 20th century when their family started Hanover Canning, which later became Hanover Foods. His family owns 2,200 acres of farmland just south of our state's capitol in Harrisburg. For decades, they've leased their land to a conventional farming family who are devout Mennonites, known in their community for their work ethic, humility, and deep-rooted agricultural traditions.

The land has been in Steve's family for a few generations. In 2019, the Warehime's brought together their holdings into a new enterprise, a strategic move to preserve farmland, reinvest in sustainability, and give back to the land that had long been part of their story.

That vision led them to partner with Rodale Institute to begin transitioning their farmland to organic. But making that shift required the cooperation and trust of the Mennonite family that was farming the land, who were still using the conventional methods they had been taught for years. As Steve put it, "When I first started talking to them about transitioning to organic, their arms were crossed. They were polite, but I was basically talking to myself."

Over time, though, the results started to speak for themselves. Rodale Institute worked side by side with Steve and the local family to manage the land organically, and eventually they harvested their first crop of certified organic corn.

Steve remembers the moment when everything shifted. In passing, one of the family members asked Steve what the price per bushel of the organic corn that they had recently harvested sold for at the local mill.

"When I told him the number, he was dumbfounded," Steve said. "And then he responded by saying, 'we're gonna try transitioning a few fields to organic on our own farm at home next season.'"

It was a simple moment, but it marked a profound shift in consciousness. From skepticism to openness. From tradition to transformation. From apathy to curiosity. And it's a reminder of what's possible when values, economics, and relationships align.

Sometimes change doesn't begin with sweeping policy shifts or top-down mandates.

Change starts in the soil, between families, over time, with a willingness to listen, trust, and try.

It's a story I hear *all* the time now.

It's as if the Earth is inviting us into a new relationship which is really an old relationship. To greater levels of connection.

As I think back on the countless conversations that shaped this book, what strikes me is how many of us share an affinity for this story. We've all left something behind - careers, comforts, habits - to step into the unknown of a future that we felt called to.

A calling, passion, mission, purpose.

In their journeys, I see echoes of my own.

And in my journey, I hope to inspire others to say "yes" to their calling to reconnect with the earth in whatever way they are called to do so.

That's the beauty of this movement: it reminds us that we're not alone. We're part of something much bigger, rooted in the soil, carrying us all forward. A massive movement that is pulling us towards an ancient future where we reconnect with the land and with each other.

When I began this work, I thought my story was unusual.

It's not.

But over and over again, I've met people who, like me, left behind one life to answer the call of the land. Each of us carries different reasons, but the same longing: to heal, to grow, to give. What started as my personal leap of faith is now a shared movement, with roots spreading far beyond me.

We're returning to the farm like never before. We are waking up to how disconnected we are from our food system. We are leaving our jobs, our careers, and reorienting our lives around food, whether that's growing it, distributing it, or working in advocacy, financing, or education.

Or maybe it's simply making a different choice at the grocery store or in our own backyard.

Making this shift requires courage and commitment. Regenerative systems do not offer instant returns. They require a willingness to invest patiently, to think beyond the next moment, and to cultivate rather than take. These choices often require some level of sacrifice in order to invest in our future.

Regeneration is about playing the long game. A call to live on the Earth with a longer view.

Healthy people rooted in regeneration will create healthier families, healthier societies, and a healthier world. We are waking up to the fact that we have been born into this life to steward the Earth.

So, what farm are you being asked to return to?

What land are you being called to heal?

The time is now.

The farm is HERE.

EPILOGUE

I have a friend named Cookie.

She used to work in high-end restaurants. Then she bought a farm.

The farm is eight miles outside of Atlantic City, New Jersey, one of the most food insecure cities in America.

The farm was going to be sold for a housing development.

Instead, at-risk youth get bussed to the farm from the city center to learn about farming and reconnect with nature.

The farm also grows food that is trucked into the city and sold at corner stores and bodegas.

Cookie used to work in high-end restaurants. Now she is running a farm that is feeding people that are severely undernourished.

Dr. Meagan Grega spent many years as a medical officer in the US Navy, served as a faculty member at Temple University School of Medicine, and worked in a private medical practice.

Now she runs a non-profit that teaches gardening in school classrooms and is upending school lunch programs by providing students with access to nutrient-dense foods grown on local and organic farms.

Sam was a rising star inside one of the most influential technology companies in recent history. Now he owns a regenerative organic farm that serves as a center for the community to reconnect with nature, to listen, and to heal. The farm grows hazelnut and chestnut trees, helping to reforest an entire ecosystem that was wiped out decades ago by blight.

Lily runs a certified organic farm at a regional hospital. The food grown on this 14-acre farm ends up on the plates of patients within the hospital system, helping to nurture them back to health.

My friend Evan transformed a one-acre dirt lot into a thriving community ecosystem. His farm is located on a historic property that is surrounded by urban sprawl, and now serves as a hub for southern California's ecological movement.

I never stop meeting people like you and me who are returning to the farm.

This book has been my attempt to tell you what I see.

Sometimes this feels like a movement, a call to arms, a growing awareness, a shift in consciousness, a change in diet, a subversion of a system.

Day after day, I never stop getting caught up in this movement.

Whatever this movement is, it's happening all over the place.

The farm is living proof that choices made for tomorrow can make a better world.

The farm isn't a distant dream.

It's here.

RODALE INSTITUTE

Named one of *Fast Company's Most Innovative Companies* in 2025, Rodale Institute is a nonprofit dedicated to growing the regenerative organic agriculture movement through rigorous research, farmer training, and education. Over its 78-year history, Rodale Institute has proven that organic farming is not only viable but essential to humanity's survival. The Institute's groundbreaking science and direct farmer support programs serve as a catalyst for change in farming and food production worldwide.

JOIN THE MOVEMENT

If this book planted a seed, here's where to help it grow:

rodaleinstitute.org/the-farm-is-here

NOTES

Chapter 1

1. Masoud Hashemi et al., "Healthy Soils," University of Massachusetts Amherst, Center for Agriculture, Food, and the Environment, UMass Extension Crops, Dairy, Livestock and Equine Program, accessed September 2, 2025, umass.edu/agriculture-food-environment/crops-dairy-livestock-equine/fact-sheets/healthy-soils.

2. Mark A. Anthony et al., "Enumerating Soil Biodiversity," *PNAS* 120, no. 33 (2023): e2304663120, doi.org/10.1073/pnas.2304663120.

3. Keith Mulvihill, "Soil Erosion 101," NRDC, June 1, 2021, https://www.nrdc.org/stories/soil-erosion-101.

4. Nicole Redvers et al., "Indigenous Peoples: Traditional Knowledges, Climate Change, and Health," *PLOS Global Public Health* 3, no. 10 (2023): e0002474, doi.org/10.1371/journal.pgph.0002474.

5. Matthew McGough et al., "How Has U.S. Spending on Healthcare Changed Over Time?" Peterson-KFF Health System Tracker, December 20, 2024, healthsystemtracker.org/chart-collection/u-s-spending-healthcare-changed-time Centers for Medicare & Medicaid Services (CMS), *National Health Expenditure Data: Historical* (U.S. Department of Health and Human Services, 2024), www.cms.gov/data-research/statistics-trends-and-reports/national-health-expenditure-data/historical; and U.S. Bureau of Economic Analysis (BEA), *Personal Consumption Expenditures by Major Type of Product: Food (DPCERG3A086NBEA)* (U.S. Department of Commerce, 2024), fred.stlouisfed.org/series/DPCERG3A086NBEA.

6. K. L. Bassil et al., "Cancer Health Effects of Pesticides: Systematic Review," *Canadian Family Physician* 53, no. 10 (2007): 1704–11, PubMed Central, PMID: 17934034; PMCID: PMC2231435.

Chapter 3

1. Rodale Institute, "Farming Systems Trial," *Rodale Institute* (website), accessed September 2, 2025, Rodale Institute, rodaleinstitute.org/science/farming-systems-trial. Background and findings of the Farming Systems Trial.

2. "Healthy Soils Are: Full of Life," USDA Natural Resources Conservation Service, accessed September 2, 2025, nrcs.usda.gov/sites/default/files/2023-01/Healthy-Soils-Are-full-of-life.pdf

3. Fuji Wang et al., "The Bacterial and Fungal Compositions in the Rhizosphere of Asarum *heterotropoides* Fr. Schmidt var. *mandshuricum* (Maxim.) Kitag. in a Typical Planting Region" *Microorganisms*, 12 no.4 (2024):692, doi.org/10.3390/microorganisms12040692.

4. Alexandra Bot and José Benites, *The Importance of Soil Organic Matter: Key to Drought-Resistant Soil and Sustained Food Production*, FAO Soils Bulletin 80 (Food and Agriculture Organization of the United Nations, Rome, 2005), 5, fao.org/4/a0100e/a0100e.pdf.

5. Don Comis, " 'Glue' That Makes Good Soil Found," *AgResearch* (October 2, 1997), USDA Agricultural Research Service, ars.usda.gov/news-events/news/research-news/1997/glue-that-makes-good-soil-found.

6. Nannipieri, Paolo. 2020. "Soil Is Still an Unknown Biological System" Applied Sciences 10, no. 11: 3717. doi.org/10.3390/app10113717.

7. National Organic Coalition, "Key Challenges to Growing the Organic Movement," Rodale Institute, January 19, 2017, rodaleinstitute.org/blog/key-challenges-to-growing-the-organic-movement.

8. "Soil Erosion and Degradation," World Wildlife Fund, accessed September 2, 2025, worldwildlife.org/threats/soil-erosion-and-degradation.

9. Crowell, Rachel. "More than 57 Billion Tons of Soil Have Eroded in the U.S. Midwest." Science News, April 12, 2022. sciencenews.org/article/soil-erosion-rate-us-midwest-unsustainable-usda.

10. Hartmann, M., B. Frey, J. Mayer, P. Mäder, and F. Widmer. "Distinct Soil Microbial Diversity under Long-Term Organic and Conventional Farming." *ISME Journal* 9, no. 5 (2015): 1177–94. doi.org/10.1038/ismej.2014.210.

11. Food and Agriculture Organization of the United Nations. *FAO in the 21st Century*. Rome: FAO, 2011. Accessed November 16, 2025. fao.org/4/i2307e/i2307e.pdf.

12. An, Yuxing, Yinglin Lu, and Guohua Zhong. "Soil Microorganisms: Their Role in Enhancing Crop Nutrition and Health." *Diversity* 16, no. 12 (2024): Article 734. doi.org/10.3390/d16120734.

13. Save Soil. "95% of the Earth's Soil on Course to Be Degraded by 2050." *Earth.Org*, June 17, 2024. https://earth.org/95-of-the-earths-soil-on-course-to-be-degraded-by-2050/.

14. International Food Policy Research Institute (IFPRI). "Hidden Hunger, Exposed." IFPRI, accessed November 16, 2025. ifpri.org/blog/hidden-hunger-exposed.

15. Tilman, David. "Soil Depletion and Nutrition Loss." *Scientific American*, November 1, 2004. scientificamerican.com/article/soil-depletion-and-nutrition-loss/.

16. Davis, Donald R., Melvin D. Epp, and Hugh D. Riordan. "Changes in USDA Food Composition Data for 43 Garden Crops, 1950 to 1999." *Journal of the American College of Nutrition* 23, no. 6 (2004): 669–82. doi.org/10.1080/07315724.2004.10719409.

17. Micha, Renata, Peilin Pei, Terry Kit, et al. "Association Between Dietary Factors and Mortality From Heart Disease, Stroke, and Type 2 Diabetes in the United States." *JAMA* 317, no. 9 (2017): 912–24. doi.org/10.1001/jama.2017.0947.

18. Centers for Disease Control and Prevention (CDC). "Childhood Obesity Facts." Last reviewed March 9, 2024. cdc.gov/obesity/childhood-obesity-facts/childhood-obesity-facts.html; "Managing Obesity in Schools." Last modified July 8, 2024. cdc.gov/school-health conditions/chronic/obesity.html.

19. Martin, S. S., A. W. Aday, N. B. Allen, et al. "2025 Heart Disease and Stroke Statistics: A Report of U.S. and Global Data From the American Heart Association." *Circulation* 151 (2025): e41-e660. doi.org/10.1161/CIR.0000000000001303.

20. Centers for Disease Control and Prevention. "Heart Disease Facts." Last updated October 24, 2024. cdc.gov/heart-disease/data-research/facts-stats/index.html.

21. GBD 2023 Cancer Collaborators. "The Global, Regional, and National Burden of Cancer, 1990-2023, with Forecasts to 2050: A Systematic Analysis for the Global Burden of Disease Study 2023." *The Lancet* 406, no. 10512 (2025): 1565-1586. doi.org/10.1016/S0140-6736(25)01635-6

22. Hagai Levine et al., "Temporal Trends in Sperm Count: A Systematic Review and Meta-Regression Analysis," *Human Reproduction Update* 23, no. 6 (2017): 646–59, https://doi.org/10.1093/humupd/dmx022.; Giulioni, C., J. M. Mundt, and F. Z. Jia. "The Environmental and Occupational Influence of Pesticides on Male Fertility: A Systematic Review of Human Studies." *Andrology* 10, no. 4 (2022): 1250–71. doi.org/10.1111/andr.13228.

23. U.S. Department of Health, Education, and Welfare, Public Health Service, National Center for Health Statistics. *Series 10, Number 51ac – Chronic Conditions: Persons 55 Years of Age and Over with One or More, Two or More, or Three or More of Six Selected Chronic Conditions, United States, 2008*. Washington, DC: U.S. Government Printing Office, 2009. cdc.gov/nchs/data/series/sr_10/sr10_051acc.pdf.

24. Centers for Disease Control and Prevention. "About Chronic Diseases." Last modified October 4, 2024. cdc.gov/chronic-disease/about/index.html.

Chapter 4

1. "Fertilizer Explodes," Wessels Living History Farm, accessed September 2, 2025, livinghistoryfarm.org/farming-in-the-1940s/crops/fertilizer-explodes.

2. Medina, Miles, Christine Angelini, et al. "Nitrogen-Enriched Discharges from a Highly Managed Watershed Intensify Red Tide (Karenia brevis) Blooms in Southwest Florida." *Science of the Total Environment* 838 (2022). https://doi.org/10.1016/j.scitotenv.2022.154149. University of Florida. "Human Activity 'Helped Fuel' Red Tide Events in Southwest Florida, New Study Reveals." Phys.org, April 8, 2022. phys.org/news/2022-04-human-fuel-red-tide-events.html

3. Rodale Institute. "Farming Systems Trial 40-Year Report." Accessed September 2, 2025. rodaleinstitute.org/wp-

content/uploads/FST_40YearReport_RodaleInstitute-1.pdf.

4. U.S. Environmental Protection Agency (EPA), "Nutrient Pollution: The Problem," last updated March 9, 2023, epa.gov/nutrientpollution/problem; US Environmental Protection Agency, "Basic Information on Nutrient Pollution" last updated April 22, 2025, epa.gov/nutrientpollution/basic-information-nutrient-pollution.

Chapter 5

1. Cary Oshins, Frederick Michel, Pierce Louis, Tom L. Richard, and Robert Rynk, "The Composting Process," in *The Composting Handbook*, ed. Robert Rynk (Academic Press, 2022), doi.org/10.1016/B978-0-323-85602-7.00008-X.

2. "Energy & Climate," The Fertilizer Institute, accessed September 2, 2025, tfi.org/advocacy/our-issues/energy-economic-growth/energy-climate.

3. National Oceanic and Atmospheric Administration. "Hypoxia: Dead Zones in the Gulf." *NOAA Ocean Service*, accessed November 17, 2025. https://oceanservice.noaa.gov/hazards/hypoxia/

4. Mikaela Conley, "Atrazine, an Endocrine Disrupting Herbicide Banned in Europe, Is Widely Used in the U.S.," U.S. Right to Know, January 7, 2025, usrtk.org/pesticides/atrazine.

5. "Earth Overshoot Day 2024 Approaching," Global Footprint Network, July 21, 2024, footprintnetwork.org/2024/07/21/earth_overshoot_day_2024.

6. "National Overview: Facts and Figures on Materials, Wastes and Recycling," Environmental Protection Agency, last modifed September 11, 2025, epa.gov/facts-and-figures-about-materials-waste-and-recycling/national-overview-

facts-and-figures-materials.

7. "World Squanders Over 1 Billion Meals a Day: UN Report," UN Environment Programme, March 27, 2024, unep.org/news-and-stories/press-release/world-squanders-over-1-billion-meals-day-un-report.

8. U.S. Environmental Protection Agency. "Composting." Last updated April 18, 2024. https://www.epa.gov/sustainable-management-food/composting

Chapter 6
1. NASA / JPL. "NASA's Kepler Discovery Confirms First Planet Orbiting Two Stars." *NASA Jet Propulsion Laboratory*, September 15, 2011. https://www.jpl.nasa.gov/news/nasas-kepler-discovery-confirms-first-planet-orbiting-two-stars/.

2. Mark A. Anthony et al., "Enumerating Soil Biodiversity," *PNAS* 120, no. 33 (2023): 1, e2304663120, doi.org/10.1073/pnas.2304663120.

3. Food and Agriculture Organization of the United Nations (FAO), *Soils, Where Food Begins: Proceedings of the Global Symposium on Soils for Nutrition, 26–29 July 2022* (FAO, Rome, 2023): 311, doi.org/10.4060/cc6728en.

4. Eric C. Brevik and Lynn C. Burgess, "The Influence of Soils on Human Health," *Nature Education Knowledge* 5, no. 12 (2014): 1, nature.com/scitable/knowledge/library/the-influence-of-soils-on-human-health-127878980.

5. Heribert Hirt, "Healthy Soils for Healthy Humans: How Beneficial Microbes in the Soil, Food and Gut Are Interconnected and How Agriculture Can Contribute to Human Health," *EMBO Reports* 21, no. 8 (2020): e51069, doi.org/10.15252/embr.202051069.

6. Rodrigo Mendes and Jos M. Raaijmakers, "Cross-Kingdom Similarities in Microbiome Functions," *The ISME Journal* 9, no. 9 (2015): 1905–1907, doi.org/10.1038/ismej.2015.7.

7. Berg, Gabriele, Gerardo V. Toledo, Jasper Schierstaedt, Heikki Hyöty, and Wisnu Adi Wicaksono. "Linking the Edible Plant Microbiome and Human Gut Microbiome." *Gut Microbes* 17, no. 1 (2025): 2551113. doi.org/10.1080/19490976.2025.2551113.

8. Rodale Institute and The Plantrician Project, *The Power of the Plate: The Case for Regenerative Organic Agriculture in Improving Human Health* (Kutztown, PA: Rodale Institute, 2019), rodaleinstitute.org/wp-content/uploads/The-Power-of-the-Plate_-Fact-Sheet_Rodale-Institute.pdf

9. Kathleen B. Watson et al.,Trends in Multiple Chronic Conditions Among US Adults, by Life Stage, Behavioral Risk Factor Surveillance System, 2013–2023, *Preventing Chronic Disease* 22 (2025): 240539, dx.doi.org/10.5888/pcd22.240539.

10. Substance Abuse and Mental Health Services Administration, *Key Substance Use and Mental Health Indicators in the United States: Results from the 2024 National Survey on Drug Use and Health* (HHS Publication No. PEP25-07-007, NSDUH Series H-60; Center for Behavioral Health Statistics and Quality, Substance Abuse and Mental Health Services Administration; 2025), samhsa.gov/data/sites/default/files/reports/rpt56287/2024-nsduh-annual-national-report.pdf

11. Chang, E. and S. Delzell. "Systematic Review and Meta-Analysis of Glyphosate Exposure and Risk of Lymphohematopoietic Cancers." *Environmental Research* (2016?): 1-9. pubmed.ncbi.nlm.nih.gov/27015139/.

12. Rodale Institute. "Crop Rotations." *Rodale Institute*. Accessed November 17, 2025. rodaleinstitute.org/why-organic/organic-farming-practices/crop-rotations/.

13. Fresh Food Farmacy: Using Food as Medicine to Manage and Prevent Diet-Responsive Conditions," Geisinger Health, accessed September 2, 2025, geisinger.org/freshfoodfarmacy.

14. Hawaii Health Matters. "Managing Chronic Disease in Hawaii." Accessed November 17, 2025. hawaiihealthmatters.org/promisepractice/index/view?pid=30471

Chapter 7

1. Pope Francis, "Address to the Second World Meeting of Popular Movements," Expo Feria Exhibition Centre, Santa Cruz de la Sierra, Bolivia, July 9, 2015, transcript quoted in Time, July 10, 2015, time.com/3952885/pope-francis-bolivia-poverty-speech-transcript.

2. Republic of Ecuador National Assembly, Legislative and Oversight Committee, *Constitution of the Republic of Ecuador*, published in the official register October 20, 2008 (Political Database of the Americas, Georgetown University, accessed September 2, 2025), Title II, Chapter 7, Articles 71–74, pdba.georgetown.edu/Constitutions/Ecuador/english08.html.

3. Shanthi Van Zeebroeck, "Nature Rights: What Countries Grant Legal Personhood Status to Nature and Why?" Earth.Org, October 6, 2022, earth.org/nature-rights.

4. Andrew Adamatzky, "Language of Fungi Derived from Their Electrical Spiking Activity," *Royal Society Open Science* 9 (2022): 211926, doi.org/10.1098/rsos.211926.

5. Richard Grant, "Do Trees Talk to Each Other?" *Smithsonian* magazine, March 2018, smithsonianmag.com/science-nature/the-whispering-trees-180968084.

6. Monica Gagliano et al., Tuned In: Plant Roots Use Sound to Locate Water, *Oecologia*, 184 (2017): 151–60, doi.org/10.1007/s00442-017-3862-z; H. M. Appel and R. B. Cocroft, "Plants Respond to Leaf Vibrations Caused by Insect Herbivore Chewing," *Oecologia* 175 (2014): 1257–66, doi.org/10.1007/s00442-014-2995-6.

7. Karl Zinsmeister, "Statistics on U.S. Generosity," in *The Almanac of American Philanthropy* (Philanthropy Roundtable, 2016), philanthropyroundtable.org/almanac/statistics-on-u-s-generosity.

8. Indiana University Lilly Family School of Philanthropy, *Giving USA 2025: The Annual Report on Philanthropy for the Year 2024* (Giving USA Foundation, 2025), available online at givingusa.org

9. Elizabeth W. Dunn et al., "Spending Money on Others Promotes Happiness, *Science* 319 (2008): 1687–88, doi.org/10.1126/science.1150952.

Chapter 8

1. Natural Resources Defense Council, *Intolerable Risk: Pesticides in Our Children's Food* (New York: NRDC, 1989).

2. Meryl Streep, testimony before the U.S. Senate Committee on Labor and Human Resources, regarding pesticide residues and children's health, 101st Cong., 1st sess., 1989, Congressional Record.

3. Gabriela Leslie (with research by Tina Owens), *Summary: Regenerative Ag CPG Targets—Mapping the Landscape of Corporate Engagement* (CREO, May 2024), 3,

dropbox.com/scl/fi/78xpnve8fjh9z9shm9qsj/2024_CREO_Su
mmary_Regenerative_Ag_CPG_Targets.pdf?rlkey=fwdjhdoh
vcrhablqg7vg8opum&e=2&st=opwiuvla&dl=0.

4. Andrew Smith, "New Report Identifies 'Toxic' Impact of No-Till Agriculture, Inaccurately Referred to As 'Regenerative,'" Rodale Institute, May, 5, 2025, rodaleinstitute.org/blog/new-report-identifies-toxic-impact-of-no-till-agriculture-inaccurately-referred-to-as-regenerative.

5. Regenerative Organic Alliance. *Regenerative Organic Certified Product Directory*. Accessed November 17, 2025. regenorganic.org/product-directory/

6. United States Department of Agriculture, Economic Research Service. "Direct-to-Consumer Farm Sales Reach $10.7 Billion in 2020." *ERS Charts of Note*. Last modified August 10, 2021. ers.usda.gov/data-products/charts-of-note/chart-detail/?chartId=104408.

7. National Gardening Association. "New Survey Shows Americans Increasingly Gardening to Benefit..." *National Wildlife Federation*, May 5, 2020. nwf.org/Latest-News/Press-Releases/2020/05-05-20-2020-National-Gardening-Survey.

8. U.S. Department of Agriculture, "USDA Invests $5.2 Million in 17 Urban Agriculture and Innovative Production Projects," July 2, 2024, fsa.usda.gov/news-events/news/07-02-2024/usda-invests-52-million-17-urban-agriculture-innovative-production.

Chapter 9

1. United States Department of Agriculture, Economic Research Service. *Pesticide Use in U.S. Agriculture: 21 Selected Crops, 1960–2008*. Economic Information Bulletin No. 124. Washington, DC: U.S. Department of Agriculture, 2014.

ers.usda.gov/sites/default/files/_laserfiche/publications/43854/46734_eib124.pdf.

2. U.S. Geological Survey. "U.S. Farms Used 544 Million Kilograms (1.2 Billion Pounds) of Pesticides in 2016." *Science News*, October 18, 2016. snexplores.org/article/us-farmers-still-use-many-pesticides-are-banned-elsewhere.

3. U.S. Geological Survey. "Herbicide Glyphosate Prevalent in U.S. Streams and Rivers." Last modified May 2, 2018. usgs.gov/news/herbicide-glyphosate-prevalent-us-streams-and-rivers.

4. Centers for Disease Control and Prevention (CDC), *Fourth National Report on Human Exposure to Environmental Chemicals*, Updated Tables (Atlanta: U.S. Department of Health and Human Services, 2019), section on Atrazine and Atrazine Metabolites, cdc.gov/exposurereport/

5. Tanner, Carla M., and Deborah J. T. Watts. "Pesticides, Parkinson's Disease, and Neurodegeneration." *Current Neurology and Neuroscience Reports* 18, no. 12 (2018): 85. doi.org/10.1007/s11910-018-0891-3.

6. Reganold, John. "Organic Profitability Around the World." *United States Department of Agriculture*. Accessed September 2, 2025. usda.gov/sites/default/files/documents/Reganold.pdf.

7. Organic Trade Association. "Research Shows Organic Agriculture Boosts Local Economies." Last modified June 2, 2016. ota.com/about-ota/press-releases/research-shows-organic-agriculture-boosts-local-economies

ACKNOWLEDGEMENTS

I cannot imagine this book existing without the help of my friends, Caitlin Elizabeth, Brent French, and Rob Bell.

Rob, thank you for showing me that I have something to say and for encouraging me to write this book.

Brent, thank you for bringing this book to life through your creative genius and expression. Your art will deeply impact this movement.

Caitlin, thank you for your edits, insights, interviews, and keeping me on track every step of the way.

Mom and Dad, thank you for providing me with the foundations of faith, health, and connection to nature. These have been the bedrock of my life.

To the Rodale Family, thank you for being the fire carriers of this mission for four generations.

To Maria Rodale and the Rodale Institute Board of Directors, thank you for your tireless and selfless service to our mission.

To the amazing and innovative team that works on the mission every day, you are making a better world possible.

To everyone that funds this mission, you are investing in a legacy that you will be proud to hand to the next generation.

To Donovan and Emily Mattole, for your extreme generosity towards this idea.

To my friends at Bloom Farm, thank you for providing me with the most inspiring space to write this book.

And to Jackie, thank you for your unwavering love, support, and constant encouragement to keep going. I love you.

ABOUT THE AUTHOR

Jeff Tkach is the CEO of Rodale Institute, the global leader in regenerative organic agriculture. His work has been featured in *The New York Times, Fast Company, Forbes,* and *Newsweek*.

Jeff lives in Pennsylvania with his wife, Jackie. When he's not working on the mission, you can find him in his garden or surfing his favorite East Coast surf breaks.

The Farm is Here is Jeff's first book.

www.ingramcontent.com/pod-product-compliance
Lightning Source LLC
LaVergne TN
LVHW010326070526
838199LV00065B/5667